REA: THE LEADER IN CLEP TEST PREP

CLEP* COLLEGE ALGEBRA

Stu Schwartz
Chestnut Hill College
Philadelphia, Pa.

Research & Education Association
Visit our website at: www.rea.com

Research & Education Association
61 Ethel Road West
Piscataway, New Jersey 08854
E-mail: info@rea.com

CLEP College Algebra with Online Practice Exams

Copyright © 2014 by Research & Education Association, Inc. Prior editions copyright © 2012, 2007, 2004, 2001, 2000, 1998, 1996. All rights reserved. No part of this book may be reproduced in any form without permission of the publisher.

Printed in the United States of America

Library of Congress Control Number 2013934984

ISBN-13: 978-0-7386-1151-8
ISBN-10: 0-7386-1151-4

All trademarks cited in this publication are the property of their respective owners.

LIMIT OF LIABILITY/DISCLAIMER OF WARRANTY: Publication of this work is for the purpose of test preparation and related use and subjects as set forth herein. While every effort has been made to achieve a work of high quality, neither Research & Education Association, Inc., nor the authors and other contributors of this work guarantee the accuracy or completeness of or assume any liability in connection with the information and opinions contained herein and in REA's software and/or online materials. REA and the authors and other contributors shall in no event be liable for any personal injury, property or other damages of any nature whatsoever, whether special, indirect, consequential or compensatory, directly or indirectly resulting from the publication, use or reliance upon this work.

Cover image © istockphoto.com/alengo

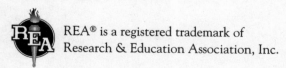

REA® is a registered trademark of Research & Education Association, Inc.

CONTENTS

About Our Author vi
A Note from Our Author vii
About Research & Education Association viii
Acknowledgments viii

CHAPTER 1
Passing the CLEP College Algebra Exam 3
    Getting Started 3
    The REA Study Center 4
    An Overview of the Exam 5
    All About the CLEP Program 6
    CLEP Exams Migrating to iBT 6
    Options for Military Personnel and Veterans 8
    SSD Accommodations for Candidates with Disabilities 9
    6-Week Study Plan 9
    In the Days Leading up to the Exam 10
    Test-Taking Tips 10
    The Day of the Exam 11
Online Diagnostic Test .... *www.rea.com/studycenter*

CHAPTER 2
Review of Math Essentials 15
    Sets 15
    Operations with Signed Numbers 19
    Powers and Roots 24
    Order of Operations 29
    Operations with Fractions 32
    Scientific Notation 37

CHAPTER 3
Algebraic Operations 45
    Operations with Algebraic Expressions 45
    Operations with Exponents 51
    Expanding and Factoring Polynomials 57
    Operations with Rational Expressions 66
    Solving Linear Equations and Inequalities 72

CHAPTER 4
Functions and Their Properties **81**
    Relations and Graphs 81
    Function Definition and Interpretation 84
    Linear Functions 91
    Graphs of Common Functions 101
    Transformation of Graphs 103
    Polynomial Graphs 108
    Inverse Functions 117

CHAPTER 5
Equations and Inequalities **123**
    Absolute Value Equations and Inequalities 123
    Quadratic Equations and Inequalities 125
    Systems of Linear Equations and Inequalities 135
    Exponential Functions and Logarithmic Equations 145

CHAPTER 6
Number Systems and Operations **159**
    Real Numbers 159
    Complex Numbers 165
    Sequences and Series 174
    Factorials and the Binomial Theorem 189
    Matrices 203

CHAPTER 7
Attacking the CLEP College Algebra Exam **209**
    Classifying CLEP Problems 209
    Sample Problems 212
    Formulas You Must Know 226
    When You See the Words… 231
    Common Mistakes to Avoid 240
    Review of Test-Taking Tips 245

Practice Test 1 (also available online at *www.rea.com/studycenter*) **247**
    Answer Key 271
    Detailed Explanations of Answers.. 273

Practice Test 2 (also available online at *www.rea.com/studycenter*) **289**
    Answer Key 313
    Detailed Explanations of Answers.. 315

Answer Sheets.. **331**

Glossary **333**

Index **339**

ABOUT OUR AUTHOR

Stu Schwartz has been teaching mathematics since 1973. For 35 years he taught in the Wissahickon School District, in Ambler, Pennsylvania, specializing in AP Calculus AB and BC and AP Statistics. He currently teaches statistics at Chestnut Hill College in Philadelphia. Mr. Schwartz received his B.S. degree in Mathematics from Temple University, Philadelphia.

Mr. Schwartz was a 2002 recipient of the Presidential Award for Excellence in Mathematics Teaching and also won the 2007 Outstanding Educator of the Year Award for the Wissahickon School District.

Outside the classroom, Mr. Schwartz does consulting work with the Math and Science Partnership of Greater Philadelphia, focusing on factors affecting success in college in STEM (science, technology, engineering, math) degrees. He is well-known in the AP community, having developed the website *www.mastermathmentor.com*, a website geared toward helping educators teach AP Calculus, AP Statistics, and other math courses. Mr. Schwartz is always looking for ways to provide teachers with new and innovative teaching materials, believing that it should be the goal of every math teacher not only to teach students mathematics, but also to find joy and beauty in math as well.

Mr. Schwartz lives with his cat, Newton, in suburban Philadelphia. He loves to bike and play piano in restaurants.

Thanks to my special friend and business partner, Ted Tyree, for always inspiring me and helping me feel that retiring provided me with an opportunity to share my enthusiasm for learning mathematics with the world.

A NOTE FROM OUR AUTHOR

I wish you great success as you embark on the task of passing the CLEP College Algebra exam. Let me offer a few words of advice on how to best use this book, which is designed to be a comprehensive guide to assist you in reaching that goal.

For many students, learning math is like going to the dentist—it isn't a fun experience. So many people are challenged by math that passing a comprehensive exam like the CLEP College Algebra exam can seem like climbing Mt. Everest.

This book was created for students of all abilities, for those who are challenged by algebra, as well as those who just want to brush up on their skills, and for all levels in between.

To maximize your score, be honest with yourself about your strengths and weaknesses in your knowledge of algebra. How strong or weak are your basic algebra skills? If all those algebra rules for simplifying expressions, working with fractions, solving equations, and graphing curves are puzzling to you, start right at the beginning of Chapter 2, which was written just for you.

Then study all the topics in Chapters 3 through 6. Each topic has example problems followed by the solutions. Test yourself by working the example problems before looking at the solutions. Do them over and over—you will be amazed how it will all start to make sense.

Once you finish the math topics, carefully read Chapter 7, which condenses all the important information in the book in several different ways, including methods for tackling problems, vital formulas to memorize, how to identify key problem types, and common pitfalls that you can avoid.

Provided are two full-length CLEP College Algebra practice exams to test your readiness for the real thing. Every problem has a solution and a classification so you can identify the problem types that you have mastered and those that are still a challenge to you.

Everything you need to pass the CLEP College Algebra exam is in these pages. As you work through this book, you will soon realize that you are learning the important concepts and developing confidence in your abilities. Who knows, it might even be fun!

Best wishes,

Stu Schwartz

ABOUT RESEARCH & EDUCATION ASSOCIATION

Founded in 1959, Research & Education Association (REA) is dedicated to publishing the finest and most effective educational materials—including study guides and test preps—for students in middle school, high school, college, graduate school, and beyond.

Today, REA's wide-ranging catalog is a leading resource for teachers, students, and professionals. Visit *www.rea.com* to see a complete listing of all our titles.

ACKNOWLEDGMENTS

We would like to thank Pam Weston, Publisher, for setting the quality standards for production integrity and managing the publication to completion; John Paul Cording, Vice President, Technology, for coordinating the design and development of the REA Study Center; Larry B. Kling, Vice President, Editorial, for his supervision of revisions and overall direction; Diane Goldschmidt, Managing Editor, for coordinating development of this edition; Kelli Wilkins for editorial review; Transcend Creative Services for typesetting this edition; and Weymouth Design and Christine Saul, Senior Graphic Designer, for designing our cover.

In addition, we extend our special thanks to Sandra Rush for her technical review and copyedit of the manuscript and Mary Willi for proofreading.

CHAPTER 1

Passing the CLEP College Algebra Exam

CHAPTER 1

PASSING THE CLEP COLLEGE ALGEBRA EXAM

Congratulations! You're joining the millions of people who have discovered the value and educational advantage offered by the College Board's College-Level Examination Program, or CLEP. This test prep focuses on what you need to know to succeed on the CLEP College Algebra exam, and will help you earn the college credit you deserve while reducing your tuition costs.

GETTING STARTED

There are many different ways to prepare for a CLEP exam. What's best for you depends on how much time you have to study and how comfortable you are with the subject matter. To score your highest, you need a system that can be customized to fit you: your schedule, your learning style, and your current level of knowledge.

This book, and the online tools that come with it, allow you to create a personalized study plan through three simple steps: assessment of your knowledge, targeted review of exam content, and reinforcement in the areas where you need the most help.

Let's get started and see how this system works.

Test Yourself and Get Feedback	Assess your strengths and weaknesses. The score report from your online diagnostic exam gives you a fast way to pinpoint what you already know and where you need to spend more time studying.
Review with the Book	Armed with your diagnostic score report, review the parts of the book where you're weak and study the answer explanations for the test questions you answered incorrectly.
Ensure You're Ready for Test Day	After you've finished reviewing with the book, take our full-length practice tests. Review your score reports and re-study any topics you missed. We give you two full-length practice tests to ensure you're confident and ready for test day.

THE REA STUDY CENTER

The best way to personalize your study plan is to get feedback on what you know and what you don't know. At the online REA Study Center, you can access two types of assessment: a diagnostic exam and full-length practice exams. Each of these tools provides true-to-format questions and delivers a detailed score report that follows the topics set by the College Board.

Diagnostic Exam

Before you begin your review with the book, take the online diagnostic exam. Use your score report to help evaluate your overall understanding of the subject, so you can focus your study on the topics where you need the most review.

Full-Length Practice Exams

These practice tests give you the most complete picture of your strengths and weaknesses. After you've finished reviewing with the book, test what you've learned by taking the first of the two online practice exams. Review your score report, then go back and study any topics you missed. Take the second practice test to ensure you have mastered the material and are ready for test day.

If you're studying and don't have Internet access, you can take the printed tests in the book. These are the same practice tests offered at the REA Study Center, but without the added benefits of timed testing conditions and diagnostic score reports. Because the actual exam is Internet-based, we recommend you take at least one practice test online to simulate test-day conditions.

AN OVERVIEW OF THE EXAM

The CLEP College Algebra exam consists of approximately 60 multiple-choice questions, each with five possible answer choices, to be answered in 90 minutes.

The exam covers the material one would find in a one-semester college-level algebra course. Test takers can expect to encounter roughly a 50/50 split between routine and nonroutine problems. Routine problems are those where students are asked to perform a specific mathematical task while nonroutine problems requires students to decide what mathematical concepts they need to solve the problem. While you will be able to avail yourself of an online scientific calculator (non-graphing, non-programmable) during the exam, none of the questions you will be asked will require its use.

The approximate breakdown of topics on the CLEP College Algebra exam is as follows:

25% Algebraic Operations
- Factoring and expanding polynomials
- Operations with algebraic expressions
- Operations with exponents
- Properties of logarithms

25% Equations and Inequalities
- Linear equations and inequalities
- Quadratic equations and inequalities
- Absolute value equations and inequalities
- Systems of equations and inequalities
- Exponential and logarithmic equations

30% Functions and Their Properties*
- Definition and interpretation
- Representation/modeling (graphical, numerical, symbolic, and verbal representations of functions)
- Domain and range
- Algebra of functions
- Graphs and their properties (including intercepts, symmetry, and transformations)
- Inverse functions

20% Number Systems and Operations
- Real numbers
- Complex numbers
- Sequences and series
- Factorials and Binomial Theorem
- Determinants of 2-by-2 matrices

ALL ABOUT THE CLEP PROGRAM

What is the CLEP?

CLEP is the most widely accepted credit-by-examination program in North America. The CLEP program's 33 exams span five subject areas. The exams assess the material commonly required in an introductory-level college course. Examinees can earn from three to twelve credits at more than 2,900 colleges and universities in the U.S. and Canada. For a complete list of the CLEP subject examinations offered, visit the College Board website: *www.collegeboard.org/clep*.

Who takes CLEP exams?

CLEP exams are typically taken by people who have acquired knowledge outside the classroom and who wish to bypass certain college courses and earn college credit. The CLEP program is designed to reward examinees for learning—no matter where or how that knowledge was acquired.

---

* Each test may contain a variety of functions, including linear, polynomial (degree ≤ 5), rational, absolute value, power, exponential, logarithmic, and piecewise-defined.

Although most CLEP examinees are adults returning to college, many graduating high school seniors, enrolled college students, military personnel, veterans, and international students take CLEP exams to earn college credit or to demonstrate their ability to perform at the college level. There are no prerequisites, such as age or educational status, for taking CLEP examinations. However, because policies on granting credits vary among colleges, you should contact the particular institution from which you wish to receive CLEP credit.

How is my CLEP score determined?

Your CLEP score is based on two calculations. First, your CLEP raw score is figured; this is just the total number of test items you answer correctly. After the test is administered, your raw score is converted to a scaled score through a process called *equating*. Equating adjusts for minor variations in difficulty across test forms and among test items, and ensures that your score accurately represents your performance on the exam regardless of when or where you take it, or on how well others perform on the same test form.

Your scaled score is the number your college will use to determine if you've performed well enough to earn college credit. Scaled scores for the CLEP exams are delivered on a 20-80 scale. Institutions can set their own scores for granting college credit, but a good passing estimate (based on recommendations from the American Council on Education) is generally a scaled score of 50, which usually requires getting roughly 66% of the questions correct.

For more information on scoring, contact the institution where you wish to be awarded the credit.

Who administers the exam?

CLEP exams are developed by the College Board, administered by Educational Testing Service (ETS), and involve the assistance of educators from throughout the United States. The test development process is designed and implemented to ensure that the content and difficulty level of the test are appropriate.

When and where is the exam given?

CLEP exams are administered year-round at more than 1,200 test centers in the United States and can be arranged for candidates abroad on request. To find the test center nearest you and to register for the exam, contact the CLEP Program:

CLEP Services
P.O. Box 6600
Princeton, NJ 08541-6600
Phone: (800) 257-9558 (8 A.M. to 6 P.M. ET)
Fax: (610) 628-3726
Website: *www.collegeboard.org*

CLEP EXAMS MIGRATING TO IBT

To improve the testing experience for both institutions and test-takers, the College Board's CLEP Program is transitioning its 33 exams from the eCBT platform to an Internet-based testing (iBT) platform. By spring 2014, all CLEP test-takers will be able to register for exams and manage their personal account information through the "My Account" feature on the CLEP website. This new feature simplifies the registration process and automatically downloads all pertinent information about the test session, making for a more streamlined check-in.

OPTIONS FOR MILITARY PERSONNEL AND VETERANS

CLEP exams are available free of charge to eligible military personnel and eligible civilian employees. All the CLEP exams are available at test centers on college campuses and military bases. Contact your Educational Services Officer or Navy College Education Specialist for more information. Visit the DANTES or College Board websites for details about CLEP opportunities for military personnel.

Eligible U.S. veterans can claim reimbursement for CLEP exams and administration fees pursuant to provisions of the Veterans Benefits Improvement Act of 2004. For details on eligibility and submitting a claim for reimbursement, visit the U.S. Department of Veterans Affairs website at *www.gibill.va.gov.*

CLEP can be used in conjunction with the Post-9/11 GI Bill, which applies to veterans returning from the Iraq and Afghanistan theaters of operation. Because the GI Bill provides tuition for up to 36 months, earning college credits with CLEP exams expedites academic progress and degree completion within the funded timeframe.

SSD ACCOMMODATIONS FOR CANDIDATES WITH DISABILITIES

Many test candidates qualify for extra time to take the CLEP exams, but you must make these arrangements in advance. For information, contact:

College Board Services for Students with Disabilities (SSD)
P.O. Box 8060
Mt. Vernon, IL 62864-0060
Phone: (609) 771-7137 (Monday through Friday, 8 A.M. to 6 P.M. ET)
TTY: (609) 882-4118
Fax: (866) 360-0114
Website: *http://student.collegeboard.org/services-for-students-with-disabilities*
E-mail: ssd@info.collegeboard.org

6-WEEK STUDY PLAN

Although our study plan is designed to be used in the six weeks before your exam, it can be condensed to three weeks by combining each two-week period into one.

Be sure to set aside enough time—at least two hours each day—to study. The more time you spend studying, the more prepared and relaxed you will feel on the day of the exam.

Week	Activity
1	Take the Diagnostic Exam at the online REA Study Center. Your score report will identify topics where you need the most review.
2–4	Study the review, focusing on the topics you missed (or were unsure of) on the Diagnostic Exam.
5	Take Practice Test 1 at the REA Study Center. Review your score report and re-study any topics you missed.
6	Take Practice Test 2 at the REA Study Center to see how much your score has improved. If you still get a few questions wrong, go back to the review and study the topics you missed.

IN THE DAYS LEADING UP TO THE EXAM

Study the second half of Chapter 7 again. It focuses on formulas that you need to know and procedures you should follow when certain types of problems appear on the exam. This section focuses on *what* to do instead of *how* to do it. You'll have increased test confidence and will maximize your score once you master this material.

TEST-TAKING TIPS

Know the format of the test. Familiarize yourself with the CLEP computer screen beforehand by logging on to the College Board website. Waiting until test day to see what it looks like in the pretest tutorial risks injecting needless anxiety into your testing experience. Also, familiarizing yourself with the directions and format of the exam will save you valuable time on the day of the actual test.

Read all the questions—completely. Make sure you understand each question before looking for the right answer. Reread the question if it doesn't make sense.

Read all of the answers to a question. Just because you think you found the correct response right away, do not assume that it's the best answer. The last answer choice might be the correct answer.

Work quickly and steadily. You will have 90 minutes to answer 60 questions, so work quickly and steadily. Taking our timed practice tests online will help you learn how to budget your time.

Use the process of elimination. Stumped by a question? Don't make a random guess. Eliminate as many of the answer choices as possible. By eliminating just two answer choices, you give yourself a better chance of getting the item correct, since there will only be three choices left from which to make your guess. Remember, your score is based only on the number of questions you answer correctly.

Answer every question. There is no penalty for guessing, so be sure to answer every question. After eliminating any obviously incorrect answer choices, answer the question to the best of your ability. If you don't have time left to review the question, there is a chance you could have guessed correctly.

Don't waste time! Don't spend too much time on any one question. Remember, your time is limited and pacing yourself is very important.

Look for clues to answers in other questions. If you guess on a question you don't know the answer to, you might find a clue to the answer elsewhere on the test.

Be sure that your answer registers before you go to the next item. Look at the screen to see that your mouse-click causes the pointer to darken the proper oval. If your answer doesn't register, you won't get credit for that question.

THE DAY OF THE EXAM

On test day, you should wake up early (after a good night's rest, of course) and have breakfast. Dress comfortably, so you are not distracted by being too hot or too cold while taking the test. (Note that "hoodies" are not allowed.) Arrive at the test center early. This will allow you to collect your thoughts and relax before the test, and it will also spare you the anxiety that comes with being late.

Before you leave for the test center, make sure you have your admission form and another form of identification, which must contain a recent photograph, your name, and signature (i.e., driver's license, student identification card, or current alien registration card). You may wear a watch. However, you may not wear one that makes noise, because it may disturb the other test-takers. No cell phones, dictionaries, textbooks, notebooks, briefcases, or packages will be permitted, and drinking, smoking, and eating are prohibited.

Good luck on the CLEP College Algebra exam!

CHAPTER 2

Review of Math Essentials

CHAPTER 2

REVIEW OF MATH ESSENTIALS

The CLEP College Algebra exam focuses on assessing students' abilities in algebra by testing concepts that should have been mastered in high school. Unfortunately, many students have never mastered some of the most basic ideas in algebra, even going back to arithmetic, and they then struggle with more advanced concepts that depend on these essential ideas.

This chapter is devoted to reviewing these most basic algebra ideas so that students will be prepared to tackle the actual subject matter of the CLEP College Algebra exam. These concepts are not tested directly on the CLEP College Algebra exam, but their mastery is vital to success on the exam. For instance, no exam problem specifically tests sets, but set notation is used in many problems on the exam. Likewise, no problem is specifically about order of operations with integers, but a number of CLEP exam problems utilize order of operations.

If you feel that your basic skills are adequate, you may still choose to look through this chapter to be sure that you are competent in all the concepts and then move on to Chapter 3. If you struggle with the basics (and be honest with yourself), spend some time with this chapter and get those essential topics straight in your mind before you move on.

SETS

A **set** is defined as a collection of objects. The objects can be related or completely unrelated. Each individual item belonging to a set is called an **element** or **member** of that set. Sets are usually represented by capital letters and are usually denoted by using braces. A typical set may be written as $A = \{2, 4, 6, 7\}$. This set has four elements. If an object k belongs to a set A, we write $k \in A$, which is read as "k is an element of set A." If k is not in set A, we write $k \notin A$. For set A then, $2 \in A$ and $5 \notin A$. Also, the order of the elements in the set makes no difference, so $A = \{2, 4, 6, 7\} = \{6, 2, 7, 4\}$.

A knowledge of several basic sets of numbers, listed below, is important. Chapter 6 expands on these sets.

1. **Natural numbers** (sometimes called counting numbers) are defined by the set $\{1, 2, 3, 4,\ldots\}$. The three dots mean that the pattern continues and, of course, this is an infinite set.

2. **Whole numbers** are defined by the set $\{0, 1, 2, 3, 4,\ldots\}$. The only difference between whole numbers and counting numbers is the inclusion of zero in the set of whole numbers.

3. **Integers** are defined by the set $\{\ldots, -3, -2, -1, 0, 1, 2, 3,\ldots\}$. Integers contain both positive and negative whole numbers. The set of integers is infinite as there is no starting point and no ending point. If an integer is positive, we do not write a sign. Therefore, $+4$ is the same as 4.

A set can be described in two ways: (1) it can be listed element by element, or (2) a rule characterizing the elements can be created. For instance, the set $K = \{1, 2, 3, 4, 5, 6, 7, 8, 9\}$ can also be described as $K = \{$natural numbers less than 10$\}$.

Sets are either finite or infinite. A **finite set** has a countable number of elements. The set $A = \{2, 4, 6, 7\}$ is finite because it has four elements. An **infinite set** is not countable. In an infinite set, it is not possible to list all the elements of the set so it must just be described.

For infinite sets, enough terms should be shown to establish the pattern before the ellipsis ($\ldots$). For example, it appears that each element in the set $P = \{1, 2, 4, 8,\ldots\}$ is doubled to find the next element.

A set not containing any elements is called the **empty set**. It is written as $\{\ \}$ or $\emptyset$. For example, $\{$natural numbers less than 1$\} = \emptyset$.

Subsets

Given two sets A and B, A is said to be a **subset** of B if every element of set A is also a member of set B. We write this as $A \subseteq B$. For example, if $A = \{1, 2, 3,$

4, 5, 6, 7, 8, 9} and $B = \{2, 4, 6, 8\}$ we could say that $B \subseteq A$ since every element in set B is also in set A. If $C = \{0, 1, 2, 3, 4\}$, it would not be true that $C \subseteq A$ because 0 is in set C and not in set A.

Several infinite sets that are subsets of the whole numbers are important:

1. {Even whole numbers} = $\{0, 2, 4, 6, 8, \ldots\}$ is the set of even whole numbers, where "even" means that the number is evenly divisible by 2. Zero is considered even because 2 divides evenly into zero.

2. {Odd whole numbers} = $\{1, 3, 5, 7, 9, \ldots\}$ is the set of odd whole numbers, where "odd" means that the number is not evenly divisible by 2.

3. {Prime numbers} = $\{2, 3, 5, 7, 11, 13, \ldots\}$ is the set of numbers that are not divisible by any number other than 1 and the number itself.

We could say that {even integers} $\subseteq$ {integers} because every even integer is also an integer.

Union and Intersection of Sets

A universal set (U) is a set from which other sets draw their elements. If A is a subset of U, then the **complement of A** (denoted as A') is the set of all elements of U that are not in A. The figure below illustrates this concept through the use of a **Venn diagram.**

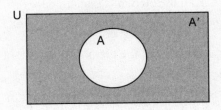

For example, if $U = \{1, 2, 3, 4, 5, 6, 7, 8, 9\}$ and $A = \{1, 3, 5, 7, 9\}$, then $A' = \{2, 4, 6, 8\}$.

The **union** of two sets A and B, denoted $A \cup B$, is the set of all elements that are either in A or in B or in both. The shaded region in the following Venn diagram illustrates this concept.

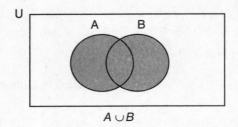

$A \cup B$

The **intersection** of two sets A and B, denoted $A \cap B$, is the set of all elements that belong to both A and B. An example is the dark-shaded region in the following Venn diagram.

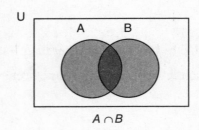

$A \cap B$

EXAMPLE

If $A = \{1, 2, 3, 4, 5, 6, 7, 8, 9\}$ and $B = \{0, 2, 5, 8, 10\}$, find $A \cup B$ and $A \cap B$.

SOLUTION

The union $A \cup B$, is all the elements that are either in A or B or both, so

$A \cup B = \{0, 1, 2, 3, 4, 5, 6, 7, 8, 9, 10\}$

The intersection, $A \cap B$, is all the elements that are in both A and B, so

$A \cap B = \{2, 5, 8\}$

If $A \cap B = \emptyset$, then A and B have no elements in common and are said to be **disjoint**. The Venn diagram in the figure below shows two disjoint sets, which have no elements in common.

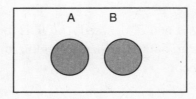

CHAPTER 2: REVIEW OF MATH ESSENTIALS

As an example, the sets of even integers and odd integers are disjoint as there is no number that is simultaneously even and odd.

EXAMPLE
If $A = \{$ all integers greater than $-2\}$, and $B = \{$all integers less than $2\}$, describe $A \cup B$ and $A \cap B$.

SOLUTION
$A = \{-1, 0, 1, 2, 3,...\}$, and $B = \{1, 0, -1, -2, -3,...\}$.

$A \cup B$ is the set of all numbers in A or B or both so
$A \cup B = \{...,-3, -2, -1, 0, 1, 2, 3,...\} = \{$integers$\}$.

$A \cap B$ is the set of all numbers common to A and B, so $A \cap B = \{-1, 0, 1\}$.

There are laws for set operations such as identity laws, complement laws, commutative laws, associative laws, distributive laws, and DeMorgan's laws. These are all outside the scope of the CLEP College Algebra exam and therefore not included in this book.

OPERATIONS WITH SIGNED NUMBERS

Working with integers means performing **operations** on them. The basic operations are adding, subtracting, multiplying, and dividing. A knowledge of how to perform these operations is absolutely vital for the CLEP College Algebra test. Let's review the rules for operations of signed numbers.

First, let's define **absolute value**. Chapters 4 and 5 will present a more formal definition of absolute value, but for now it is defined as the number without regard to sign. Absolute values will always be positive, with the exception of zero, whose absolute value is 0.

So $|5| = 5$, $|-5| = 5$, $\left|\frac{2}{3}\right| = \frac{2}{3}$, $|-7.5| = 7.5$, and $|0| = 0$.

Rules for Adding Integers

1. To add two numbers that have the same sign, add their absolute values. The sign of the answer will be the sign of the addends: a positive plus

a positive equals a positive, and a negative plus a negative equals a negative.

2. To add two numbers that have opposite signs, subtract the absolute values. The sign of the answer will be the sign of the addend with the greatest absolute value.

EXAMPLE

Find a. $8 + 2$

b. $-6 + (-7)$

c. $6 + (-10)$

d. $-8 + (-4) + 13$

SOLUTION

a. Both 8 and 2 are positive, so the answer is positive: $8 + 2 = 10$.

b. Both numbers are negative, so the answer will be negative: $-6 + (-7) = -13$.

c. The two numbers have opposite signs. Since $|-10| > |6|$, the answer will be negative, so $6 + (-10) = -4$.

d. Add the numbers in order. $-8 + (-4) = -12$, and $-12 + 13 = 1$.

Rule for Subtracting Integers

For all integers a and b, $a - b = a + (-b)$. This says that all subtraction problems should be thought of as the addition of negatives with the addition rules given above.

EXAMPLE

Find the following:

a. $8 - 5$

b. $2 - 12$

c. $3 - (-7)$

d. $-80 - (-70) - 10$

SOLUTION

a. This is the same type of subtraction learned in first grade. Technically, we are saying that $8 - 5 = 8 + (-5) = 3$.

b. $2 - 12 = 2 + (-12) = -10$.

c. To subtract -7 is the same as adding the negative of -7, which is the same as adding 7. So $3 - (-7) = 3 + 7 = 10$.

d. $-80 - (-70) - 10 = -80 + 70 + (-10) = -10 + (-10) = -20$.

By now, whenever we see a problem such as $4 - 10$, we should immediately see that the answer is -6, rather than taking the time to write it out as an addition problem. Taking the time to write the extra step is sometimes helpful, though.

Rules for Multiplying Integers

The rules for multiplying two integers are much more straightforward than the rules for addition. In algebra, we generally stay away from the multiplication signs ($\times$ or $\cdot$) and use parentheses to signal multiplication.

1. If the two integers have the same sign (either both positive or both negative), the answer is positive.

2. If the two integers have opposite signs (one positive and one negative, in either order), the answer is negative.

3. If one of the numbers is zero, the answer is zero. Zero has no sign.

EXAMPLE

Find the following:

a. $8(4)$

b. $-17(-2)$

c. $-5(9)$

d. $-8(-6)(-4)(-2)(0)(2)(4)$

SOLUTION

a. Both numbers are positive, so the answer is positive: $8(4) = 32$.

b. Both numbers are negative, so the answer is positive: $-17(-2) = 34$.

c. One number is positive and the other negative, so the answer is negative: $-5(9) = -45$.

d. When multiplying many numbers, check first to see whether one of them is zero, in which case the answer is always zero.

When multiplying many integers (other than zero), first count the number of negative integers. If there is an even number of negative integers, the result is positive. If there is an odd number of negative integers, the result is negative.

EXAMPLE

Find $2(-1)(-2)(-3)(2)(-1)(2)(-2)(-1)$.

SOLUTION

There are six negative integers, so the result is positive. Once we determine the sign, we multiply the absolute values: $2(1)(2)(3)(2)(1)(2)(2)(1) = 96$.

Rules for Dividing Integers

When we divide an integer a by an integer b, we write the problem as a fraction, so $a \div b$ is equivalent to $\frac{a}{b}$. The result of this division, called a **quotient**, may or may not be an integer. If b divides into a evenly, the quotient is an integer. If not, the quotient will be in the form of a fraction. This fraction is a **rational** number, which is discussed in more detail in Chapter 6. This rational number $\frac{a}{b}$ must be in lowest terms, meaning that there is no integer that divides evenly into both a and b. Thus, $\frac{50}{75}$ reduces to $\frac{2}{3}$ in lowest terms because 25 divides evenly into both 50 and 75. The section "Reducing Fractions" later in this chapter explains how this is done.

One of the most important rules in division is that we can never divide by zero. Any number divided by zero is said not to exist. However, zero divided by any number (except zero) is zero. So $\frac{0}{5} = 0$, $\frac{0}{-9} = 0$, but $\frac{1}{0}$ does not exist and $\frac{0}{0}$ does not exist.

In terms of signs when dividing integers, the rules are the same as for multiplying integers.

CHAPTER 2: REVIEW OF MATH ESSENTIALS | 23

1. If the two integers have the same sign (either both positive or both negative), the quotient is positive.

2. If the two integers have opposite signs (one positive and one negative, in either order), the quotient is negative.

3. Fractions that are negative can be written with the negative sign in the numerator, the denominator (less frequently done), or in front of the fraction. So $\dfrac{-a}{b} = \dfrac{a}{-b} = -\dfrac{a}{b}$ if $b \neq 0$.

EXAMPLE

Find the following:

a. $\dfrac{6}{-3}$

b. $\dfrac{-4}{8}$

c. $\dfrac{-20}{-24}$

d. $\dfrac{|12-30|}{|2-6|}$

e. $-\left|\dfrac{-3(-4)}{-2-6}\right|$

SOLUTION

a. One integer is positive and the other is negative, so the quotient is negative: $\dfrac{6}{-3} = -2$.

b. One integer is positive and the other is negative, so the quotient is negative: $\dfrac{-4}{8} = -\dfrac{1}{2}$.

c. Both integers are negative, so the quotient is positive: $\dfrac{-20}{-24} = \dfrac{4(5)}{4(6)} = \dfrac{5}{6}$.

d. $\dfrac{|12-30|}{|2-6|} = \dfrac{|-18|}{|-4|} = \dfrac{18}{4} = \dfrac{2(9)}{2(2)} = \dfrac{9}{2}$.

e. $-\left|\dfrac{-3(-4)}{-2-6}\right| = -\left|\dfrac{12}{-8}\right| = -\dfrac{12}{8} = -\dfrac{4(3)}{4(2)} = -\dfrac{3}{2}$.

POWERS AND ROOTS

Powers

Raising an integer to a whole number power is a convenient way to write a multiplication problem. So a^b is defined as $a^b = \underbrace{a(a)(a)...(a)}_{b \text{ factors}}$, where a is called the **base** and b is called the **power** or **exponent**. Thus, $2^4 = 2(2)(2)(2) = 16$. Note that 2^4 is not the same as $2(4)$.

When we raise any integer (except zero) to the zero power, the answer is 1; thus, $3^0 = 1$. We also use as a convention that any integer raised to the first power is the integer itself, and there is no need to write the power; for example, $3^1 = 3$.

EXAMPLE

Find the following:

a. 3^3

b. $(-5)^4$

c. $\dfrac{(-4)^4}{(-2)^3}$

SOLUTION

a. $3^3 = 3(3)(3) = 27$. (Don't fall into the trap of thinking that $3^3 = 9$.)

b. $(-5)^4 = -5(-5)(-5)(-5) = 625$.

c. $\dfrac{(-4)^4}{(-2)^3} = \dfrac{-4(-4)(-4)(-4)}{-2(-2)(-2)} = \dfrac{256}{-8} = -32$.

When we raise negative numbers to a large power, we are frequently interested in the sign of the answer. If the exponent is even, the answer will be positive, and if the exponent is odd, the answer will be negative.

EXAMPLE

Find the signs of the following:

a. 2^{19}
b. $(-2)^{19}$
c. $(-1)^{100}$
d. $(-5)^0$

SOLUTION

a. Since 2 is positive, we are multiplying 19 positive numbers and the result is positive.
b. Since 19 is odd, we are multiplying a negative number an odd number of times, so the result is negative.
c. Since 100 is even, we are multiplying a negative number an even number of times and the result is positive.
d. *Any* integer (except zero) raised to the zero power is 1, so the answer is positive.

In algebra, a frequent process is **squaring** numbers, that is, raising numbers to the second power. **Cubing** numbers, raising them to the third power, and raising numbers to the fourth power happen less frequently. The following table summarizes results that should be readily recognized.

Powers

Base	Power 2	Power 3	Power 4
0	$0^2 = 0$	$0^3 = 0$	$0^4 = 0$
1	$1^2 = 1$	$1^3 = 1$	$1^4 = 1$
2	$2^2 = 4$	$2^3 = 8$	$2^4 = 16$
3	$3^2 = 9$	$3^3 = 27$	$3^4 = 81$
4	$4^2 = 16$	$4^3 = 64$	$4^4 = 256$
5	$5^2 = 25$	$5^3 = 125$	$5^4 = 625$
6	$6^2 = 36$		
7	$7^2 = 49$		

Powers (cont'd)

Base	Power 2	Power 3	Power 4
8	$8^2 = 64$		
9	$9^2 = 81$		
10	$10^2 = 100$	$10^3 = 1000$	$10^4 = 10{,}000$
11	$11^2 = 121$		
12	$12^2 = 144$		
15	$15^2 = 225$		
20	$20^2 = 400$		
25	$25^2 = 625$		

The results in the Power 2 column are called **perfect squares**, and the results in the Power 3 column are called **perfect cubes**. For instance, 49 is a perfect square because $7^2 = 49$, and 27 is a perfect cube because $3^3 = 27$.

Roots

The reverse process of raising integers to powers is called **finding a root**. If the process is squaring, it is called **finding the square root**. The symbol for square root is $\sqrt{}$. For example, $9^2 = 81$, so we can say that $\sqrt{81} = 9$. Note that the square root of a positive number is always a positive number. It is incorrect to say that $\sqrt{16} = \pm 4$. Also realize that it is impossible to take the square root of a negative number, because there is no number that multiplied by itself will yield a negative result. The root $\sqrt{-16}$ does not exist. (Note: Chapter 6, which addresses imaginary numbers, revisits this problem.)

The reverse process of cubing is called **finding the cube root**. The symbol for cube root is $\sqrt[3]{}$. For example, since $2^3 = 8$, we can say that $\sqrt[3]{8} = 2$. Note that the cube root of a positive number is always a positive number. Also realize that, unlike square roots, we can take the cube root of negative numbers. Thus, $\sqrt[3]{-8} = -2$ because $(-2)^3 = -2(-2)(-2) = -8$.

In general, a **radical** is in the form of $\sqrt[n]{a}$, where a is an integer and n is a positive integer. The value under the radical sign, a, is called the **radicand**, and the variable n is called the **index**. Note that if $n = 2$ (a square root), we don't write the index. So $\sqrt[2]{25} = \sqrt{25} = 5$.

The following chart summarizes the results of finding roots.

	Positive radicand	Negative radicand
Even index : $\sqrt{}, \sqrt[4]{}, \sqrt[6]{}$, etc.	Positive result	Result does not exist
Odd index : $\sqrt[3]{}, \sqrt[5]{}, \sqrt[7]{}$, etc.	Positive result	Negative result

EXAMPLE

Find the following roots, if possible.

a. $\sqrt{81}$

b. $\sqrt{-81}$

c. $\sqrt[3]{125}$

d. $\sqrt[3]{-125}$

e. $\sqrt[11]{-1}$

f. $\sqrt[12]{-1}$

SOLUTION

a. $\sqrt{81} = 9$.

b. $\sqrt{-81}$ does not exist because the index is even and the radicand is negative.

c. $\sqrt[3]{125} = 5$.

d. $\sqrt[3]{-125} = -5$.

e. $\sqrt[11]{-1} = -1$.

f. $\sqrt[12]{-1}$ does not exist because the index is even and the radicand is negative.

Frequently, we wish to simplify radicals by multiplying them. We can use the fact that $\sqrt{a} \cdot \sqrt{b} = \sqrt{ab}$ when a and b are positive integers. So we can say that $\sqrt{7} \cdot \sqrt{3} = \sqrt{21}$ and $\sqrt{6} \cdot \sqrt{6} = \sqrt{36} = 6$.

Rather than take a square root of a large number, we can reduce it by taking the square roots of its factors, which can be much smaller numbers. To do so, we reverse the above multiplication: $\sqrt{ab} = \sqrt{a} \cdot \sqrt{b}$.

If we can break up the radicand into two factors and one of them is a perfect square, we can simplify the expression. For example, $\sqrt{28} = \sqrt{4(7)} = \sqrt{4} \cdot \sqrt{7} = 2\sqrt{7}$. This works with cube roots (or any other root) as well: $\sqrt[3]{16} = \sqrt[3]{8(2)} = \sqrt[3]{8}\sqrt[3]{2} = 2\sqrt[3]{2}$.

EXAMPLE

Find the following:

a. $\sqrt{45}$

b. $\sqrt{72}$

c. $\sqrt{75}$

d. $\sqrt{500}$

e. $\sqrt[3]{54}$

SOLUTION

a. $\sqrt{45} = \sqrt{9(5)} = \sqrt{9}\sqrt{5} = 3\sqrt{5}$.

b. $\sqrt{72} = \sqrt{36(2)} = \sqrt{36}\sqrt{2} = 6\sqrt{2}$.

c. $\sqrt{75} = \sqrt{25(3)} = \sqrt{25}\sqrt{3} = 5\sqrt{3}$.

d. $\sqrt{500} = \sqrt{100(5)} = \sqrt{100}\sqrt{5} = 10\sqrt{5}$.

e. $\sqrt[3]{54} = \sqrt[3]{27(2)} = \sqrt[3]{27}\sqrt[3]{2} = 3\sqrt[3]{2}$.

When we take a root of a fraction, we use the fact that $\sqrt{\dfrac{a}{b}} = \dfrac{\sqrt{a}}{\sqrt{b}}$. But that places the radical in the denominator of a fraction, and radicals are frequently not permitted in the denominator as a final answer to a problem. We can easily fix that, however, by multiplying both the numerator and the denominator by the radical, which doesn't change the value because we essentially are multiplying by 1. For example, $\sqrt{\dfrac{1}{2}} = \dfrac{\sqrt{1}}{\sqrt{2}} = \dfrac{1}{\sqrt{2}}$. Since we have a radical in the denominator, we multiply both the numerator and the denominator by $\sqrt{2}$ to get $\dfrac{1}{\sqrt{2}}\left(\dfrac{\sqrt{2}}{\sqrt{2}}\right) = \dfrac{\sqrt{2}}{\sqrt{4}} = \dfrac{\sqrt{2}}{2}$. Thus, $\sqrt{\dfrac{1}{2}} = \dfrac{\sqrt{2}}{2}$.

CHAPTER 2: REVIEW OF MATH ESSENTIALS

EXAMPLE

Find the following:

a. $\sqrt{\dfrac{7}{4}}$

b. $\sqrt{\dfrac{2}{3}}$

c. $\sqrt{\dfrac{5}{8}}$

d. $\sqrt{\dfrac{3}{20}}$

SOLUTION

a. $\sqrt{\dfrac{7}{4}} = \dfrac{\sqrt{7}}{\sqrt{4}} = \dfrac{\sqrt{7}}{2}.$

b. $\sqrt{\dfrac{2}{3}} = \dfrac{\sqrt{2}}{\sqrt{3}}\left(\dfrac{\sqrt{3}}{\sqrt{3}}\right) = \dfrac{\sqrt{6}}{\sqrt{9}} = \dfrac{\sqrt{6}}{3}.$

c. $\sqrt{\dfrac{5}{8}} = \dfrac{\sqrt{5}}{\sqrt{8}}\left(\dfrac{\sqrt{8}}{\sqrt{8}}\right) = \dfrac{\sqrt{40}}{\sqrt{64}} = \dfrac{\sqrt{4}\sqrt{10}}{8} = \dfrac{2\sqrt{10}}{8} = \dfrac{\sqrt{10}}{4}.$

d. $\sqrt{\dfrac{3}{20}} = \dfrac{\sqrt{3}}{\sqrt{20}}\left(\dfrac{\sqrt{20}}{\sqrt{20}}\right) = \dfrac{\sqrt{60}}{\sqrt{400}} = \dfrac{\sqrt{4}\sqrt{15}}{20} = \dfrac{2\sqrt{15}}{20} = \dfrac{\sqrt{15}}{10}.$

ORDER OF OPERATIONS

There are many possibilities (listed below) for finding the value of $8 - 6 \div 2 \times 3$, but only one is correct.

$8 - 6 \div 2 \times 3 = 2 \div 2 \times 3 = 1 \times 3 = 3$

$8 - 6 \div 2 \times 3 = 8 - 3 \times 3 = 5 \times 3 = 15$

$8 - 6 \div 2 \times 3 = 8 - 3 \times 3 = 8 - 9 = -1$

$8 - 6 \div 2 \times 3 = 8 - 6 \div 6 = 2 \div 6 = \dfrac{1}{3}$

$8 - 6 \div 2 \times 3 = 8 - \dfrac{6}{6} = 8 - 1 = 7$

When encountering an expression that has several operations, it is of vital importance that we develop an **order of operations** that need to be performed. The order of operations that is universally understood is as follows:

1. Do all operations within parentheses first.
2. Next, evaluate any part of the expression that contains exponents.
3. Then do any multiplication and/or division in order from left to right.
4. Finally, do any addition or subtraction in order from left to right.

A popular mnemonic that students memorize when learning the order of operations is "**P**lease **E**xcuse **M**y **D**ear **A**unt **S**ally," where **P** corresponds to parentheses, **E** corresponds to exponents, **M** corresponds to multiplication, **D** corresponds to division, **A** corresponds to addition, and **S** corresponds to subtraction.

EXAMPLE

Find the value of $8 - 6 \div 2 \times 3$.

SOLUTION

Since there are no parentheses or exponents, we perform the multiplications and divisions left to right. Note that multiplication is not more important than division—we do these operations as they appear, left to right.

$8 - 6 \div 2 \times 3$.
$= 8 - 3 \times 3$
$= 8 - 9$.

Finally, we do the subtraction and our answer is -1.

EXAMPLE

Find the value of $3(10 + 8) + 4^2$.

SOLUTION

$3(10 + 8) + 4^2$
$= 3(18) + 4^2$
$= 3(18) + 16$
$= 54 + 16$
$= 70$

EXAMPLE

Find the value of $(41 - 3^2) \div (12 + 2 \times 2)$.

SOLUTION

$(41 - 3^2) \div (12 + 2 \times 2)$

$= (41 - 9) \div (12 + 4)$

$= 32 \div 16$

$= 2.$

We do all of the calculations within the parentheses first, using the order of operations within each pair of parentheses, and then clear the parentheses away once we get these values.

When there are parentheses inside of parentheses (called **nested parentheses**), we work from the innermost parentheses to the outermost parentheses. Typically, brackets are used to make the problem clearer.

EXAMPLE

Find the value of $4 - \left[4^2 - \left(\dfrac{24}{3} - 5^2 \right) \right]$.

SOLUTION

$4 - \left[4^2 - \left(\dfrac{24}{3} - 5^2 \right) \right]$

$= 4 - \left[16 - \left(\dfrac{24}{3} - 25 \right) \right]$

$= 4 - [16 - (8 - 25)]$

$= 4 - [16 - (-17)]$

$= 4 - (16 + 17)$

$= 4 - 33$

$= -29$

EXAMPLE

Find the value of $6^2 \div \left(3\left[5^2 - \left(2^3 - 3^2\right)^2\right]\right)$.

SOLUTION

$$6^2 \div \left(3\left[5^2 - \left(2^3 - 3^2\right)^2\right]\right)$$
$$= 6^2 \div \left(3\left[25 - (8-9)^2\right]\right)$$
$$= 6^2 \div \left(3\left[25 - (-1)^2\right]\right)$$
$$= 6^2 \div \left(3[25 - 1]\right)$$
$$= 6^2 \div \left(3[24]\right)$$
$$= 36 \div 72$$
$$= \frac{1}{2}$$

OPERATIONS WITH FRACTIONS

Another term for fractions is **rational numbers**, which are in the form $\frac{a}{b}$, $b \neq 0$, where a and b are integers. Chapter 3 deals with rational expressions that are fractions using variables. Here, however, we concentrate on working with fractions with numerators and denominators containing integers, a topic that is taught in grade school, but one that many students have difficulty mastering.

Reducing Fractions

As stated before, fractions must be **reduced to lowest terms**. If an answer to a problem in the CLEP exam is in fraction form, it will probably be reduced. The way to reduce a fraction is to write both the numerator and the denominator as the product of two (or more integers) called **factors**. Then, we can cancel all the factors that appear in both the numerator and the denominator, as in $\frac{\cancel{a} \cdot b}{\cancel{a} \cdot c} = \frac{b}{c}$. Finally, if necessary, we multiply the remaining factors in the numerators together and those in the denominators together. It is permissible (and preferable) to leave a fraction in **improper form**, where the absolute value of the numerator is greater than the absolute value of the denominator, such as $\frac{17}{3}$.

EXAMPLE

Reduce to lowest terms:

a. $\dfrac{9}{15}$

b. $-\dfrac{24}{36}$

c. $\dfrac{-252}{-98}$

SOLUTION

a. $\dfrac{9}{15} = \dfrac{\cancel{3}(3)}{\cancel{3}(5)} = \dfrac{3}{5}.$

b. $-\dfrac{24}{36} = -\dfrac{\cancel{12}(2)}{\cancel{12}(3)} = -\dfrac{2}{3}.$

c. $\dfrac{-252}{-98} = \dfrac{\cancel{2}(126)}{\cancel{2}(49)} = \dfrac{2(\cancel{7})(9)}{\cancel{7}(7)} = \dfrac{18}{7}.$

The solution to example c could be done in only two steps if we recognize that 14 is a common factor. Note that the answer is left in improper form.

A common mistake made by students is to try to perform cancellation in the expression $\dfrac{4+5}{4}$. We know the answer is $\dfrac{9}{4}$, but canceling the 4's might lead to the (wrong) answer of $\dfrac{1+5}{1} = 6$. The only time we can cancel numbers in both the numerator and the denominator is when the operation is multiplication. If it is addition or subtraction, do not attempt to cancel.

Multiplication of Fractions

Multiplying fractions is easy. The rule states that to multiply two fractions, we multiply the numerators and multiply the denominators, and then reduce the fraction to its lowest terms, if possible: $\left(\dfrac{a}{b}\right)\left(\dfrac{c}{d}\right) = \dfrac{ac}{bd}.$

EXAMPLE

Find $\dfrac{3}{4}\left(\dfrac{5}{7}\right)$.

SOLUTION

$\dfrac{3}{4}\left(\dfrac{5}{7}\right) = \dfrac{3(5)}{4(7)} = \dfrac{15}{28}$ (which cannot be reduced).

EXAMPLE

Find $\dfrac{2}{5}\left(\dfrac{15}{14}\right)$.

SOLUTION

$\dfrac{2}{5}\left(\dfrac{15}{14}\right) = \dfrac{30}{70} = \dfrac{3}{7}$.

This problem can also be solved by canceling: 15 can be written as 5(3), and 14 can be written as 7(2). Remembering that, with multiplication, any number on the top can cancel any number on the bottom, we get $\dfrac{2}{5}\left(\dfrac{15}{14}\right) = \dfrac{\cancel{2}}{\cancel{5}}\left(\dfrac{\cancel{5}\cdot 3}{7\cdot \cancel{2}}\right) = \dfrac{3}{7}$.

EXAMPLE

Find $3\left(\dfrac{-6}{5}\right)\left(\dfrac{20}{21}\right)$.

SOLUTION

To help with the cancellation, we write 3 as $\dfrac{3}{1}$:

$\dfrac{\cancel{3}}{1}\left(\dfrac{-3\cdot 2}{\cancel{5}}\right)\left(\dfrac{\cancel{5}\cdot 4}{\cancel{3}\cdot 7}\right) = \dfrac{-24}{7}$.

Dividing Fractions

For simple fractions, we use the rule that dividing by a fraction is the same as multiplying by the fraction's reciprocal. The **reciprocal** is 1 over the fraction, which results in the fraction being turned "upside down." So $\dfrac{\frac{a}{b}}{\frac{c}{d}} = \dfrac{a}{b}\left(\dfrac{d}{c}\right) = \dfrac{ad}{bc}$, where $b, c, d \neq 0$. When the problem is more complex, as when the numerators and denominators themselves contain addition or subtraction problems, the method is different; that method is reviewed in Chapter 3.

EXAMPLE

Find $\dfrac{\frac{3}{4}}{\frac{7}{2}}$.

SOLUTION

$$\dfrac{\frac{3}{4}}{\frac{7}{2}} = \frac{3}{4}\left(\frac{2}{7}\right) = \frac{6}{28} = \frac{\cancel{2}(3)}{\cancel{2}(14)} = \frac{3}{14}.$$

EXAMPLE

Find $\dfrac{\frac{2}{9}}{-6}$.

SOLUTION

$$\dfrac{\frac{2}{9}}{-6} = \frac{2}{9}\left(\frac{-1}{6}\right) = \frac{-2}{54} = \frac{-1}{27}.$$

EXAMPLE

Find $\dfrac{-\frac{2}{9}\left(\frac{3}{5}\right)}{\frac{8}{15}\left(-\frac{4}{3}\right)}$.

SOLUTION

$$\dfrac{-\frac{2}{9}\left(\frac{3}{5}\right)}{\frac{8}{15}\left(-\frac{4}{3}\right)} = -\frac{2}{9}\left(\frac{3}{5}\right)\left(\frac{15}{8}\right)\left(-\frac{3}{4}\right)$$

$$= -\frac{\cancel{2}}{\cancel{3}(\cancel{3})}\left(\frac{\cancel{3}}{\cancel{5}}\right)\left(\frac{\cancel{3}(\cancel{5})}{\cancel{2}(4)}\right)\left(-\frac{3}{4}\right) = \frac{3}{16}.$$

Adding and Subtracting Fractions

Adding or subtracting fractions is not as straightforward as multiplying or dividing fractions. The two rules to remember are:

1. To add or subtract two or more fractions, we need a common denominator. The denominators must be the same for all fractions. If they are the same, simply add (or subtract) the numerators. Symbolically, $\dfrac{a}{c} + \dfrac{b}{c} = \dfrac{a+b}{c}, c \neq 0$.

2. If the denominators are not the same, we must do some work to get them to be the same. We need to find a **common denominator (CD)**, which is a number into which both denominators divide evenly. There are many possible common denominators; the simplest way to find a common denominator is to multiply both denominators together.

Once we find the common denominator, we need to write each fraction with that common denominator. Symbolically, if we wish to add $\dfrac{a}{b} + \dfrac{c}{d}$, we use the common denominator bd.

To write the first fraction $\dfrac{a}{b}$ with common denominator bd, we multiply it by $\dfrac{d}{d}$ getting $\dfrac{a}{b}\left(\dfrac{d}{d}\right)$.

To write the second fraction $\dfrac{c}{d}$ with common denominator bd, we multiply it by $\dfrac{b}{b}$, getting $\dfrac{c}{d}\left(\dfrac{b}{b}\right)$.

Finally, since the denominators are the same, we can just add the numerators: $\dfrac{a}{b} + \dfrac{c}{d} = \dfrac{a}{b}\left(\dfrac{d}{d}\right) + \dfrac{c}{d}\left(\dfrac{b}{b}\right) = \dfrac{ad+bc}{bd}$.

EXAMPLE

Find the values of the following:

a. $\dfrac{3}{4} + 2$

b. $\dfrac{5}{6} + \dfrac{2}{5}$

c. $\dfrac{4}{3} - \dfrac{5}{6}$

d. $\dfrac{7}{8} - \dfrac{1}{6}$

e. $\dfrac{2}{5} - \dfrac{9}{4} + \dfrac{3}{2}$

SOLUTION

a. CD = 4: $\quad \dfrac{3}{4} + 2 = \dfrac{3}{4} + \dfrac{2}{1}\left(\dfrac{4}{4}\right) = \dfrac{3+8}{4} = \dfrac{11}{4}.$

b. CD = 30: $\quad \dfrac{5}{6} + \dfrac{2}{5} = \dfrac{5}{6}\left(\dfrac{5}{5}\right) + \dfrac{2}{5}\left(\dfrac{6}{6}\right) = \dfrac{25+12}{30} = \dfrac{37}{30}.$

c. CD = 18: $\quad \dfrac{4}{3} - \dfrac{5}{6} = \dfrac{4}{3}\left(\dfrac{6}{6}\right) - \dfrac{5}{6}\left(\dfrac{3}{3}\right) = \dfrac{24-15}{18} = \dfrac{9}{18} = \dfrac{1}{2}.$

d. CD = 48: $\quad \dfrac{7}{8} - \dfrac{1}{6} = \dfrac{7}{8}\left(\dfrac{6}{6}\right) - \dfrac{1}{6}\left(\dfrac{8}{8}\right) = \dfrac{42-8}{48} = \dfrac{34}{48} = \dfrac{17}{24}.$

e. CD = 40: $\quad \dfrac{2}{5} - \dfrac{9}{4} + \dfrac{3}{2} = \dfrac{2}{5}\left(\dfrac{8}{8}\right) - \dfrac{9}{4}\left(\dfrac{10}{10}\right) + \dfrac{3}{2}\left(\dfrac{20}{20}\right)$

$\qquad\qquad\qquad = \dfrac{16 - 90 + 60}{40} = \dfrac{-14}{40} = \dfrac{-7}{20}.$

Note: A more effective method is finding the **lowest common denominator (LCD)**, which is the *smallest* possible common denominator. The method above also works, but the fraction may have to be reduced at the end of the problem.

SCIENTIFIC NOTATION

It is possible that some of the numbers or answers on the CLEP College Algebra exam may be in **scientific notation**. Although we use standard notation for most numbers, we use scientific notation when working with numbers that are either very large or very small. For example, the number of cells in the human body is estimated to be 150,000,000,000,000 (read as "150 trillion"), which can be written as 1.5×10^{14}. The length of time it takes light to travel 1 meter is approximately 0.000000003 second (read as "3 billionths of a second"), which is written as 3×10^{-9} seconds.

The general form for a number in scientific notation is $a \times 10^n$ where $1 \leq a < 10$ and n is an integer.

Converting Scientific Notation to Standard Notation

To convert a number in scientific notation to standard notation, we move the decimal point, adding zeros to fill in the spaces as necessary. If n is positive, we move the decimal point n places to the right. If n is negative, we move the decimal point n places to the left.

For example, to convert 1.2×10^5 to standard notation, we need to move the decimal point five places to the right. There is already one decimal place to the right of the decimal point, so we must add four zeros. Thus, $1.2 \times 10^5 = 120000$. We usually write 120000 as 120,000, making it easier to read.

EXAMPLE

Change the following numbers in scientific notation to standard notation.

a. 3.7×10^4

b. 9.92×10^2

c. 1.005×10^9

SOLUTION

a. $3.7 \times 10^4 = 37,000.$

b. $9.92 \times 10^2 = 992.$

c. $1.005 \times 10^9 = 1,005,000,000.$

To convert 4.7×10^{-4}, we need to move the decimal point four places to the left. There is already one decimal place to the left of the decimal point, so we need to add three zeros. Thus, $4.7 \times 10^{-4} = 0.00047$. It is customary to also put a zero before the decimal point in standard notation.

EXAMPLE

Change the following numbers in scientific notation to standard notation:

a. 5.1×10^{-3}

b. 2.27×10^{-6}

c. 8.004×10^{-5}

d. 4.2×10^0

SOLUTION

a. $5.1 \times 10^{-3} = 0.0051$.

b. $2.27 \times 10^{-6} = 0.00000227$.

c. $8.004 \times 10^{-5} = 0.00008004$.

d. $4.2 \times 10^0 = 4.2$.

Converting Standard Notation to Scientific Notation

To convert numbers to scientific notation, we focus on large numbers (numbers much greater than 1) or small numbers (numbers much less than 1).

Large numbers: To convert 34,320 to scientific notation, we place a caret after the first nonzero digit, the first 3. We then count the number of decimal places from the caret to the real decimal point, which is at the end of the whole number. There are four decimal places, so we move the decimal point four places to the left to the caret and use 10^4. Our answer is $34,320 = 3.432 \times 10^4$.

If the number is 25,460,000,000, as another example, we place a caret after the first nonzero digit, the 2. We then count the number of decimal places from the caret to the real decimal point, which is at the end of the whole number. There are ten decimal places, so we move the decimal point ten places to the left to the caret and use 10^{10}. Our answer is $25,460,000,000 = 2.546 \times 10^{10}$.

Small numbers: To convert 0.000073 to scientific notation, we place a caret after the first nonzero digit, the 7. We then count the number of decimal places from the caret to the original decimal point. There are five decimal places, so we move the decimal point five places to the right to the caret and use 10^{-5}. Our answer is $0.000073 = 7.3 \times 10^{-5}$.

If the number is 0.01503, as another example, we place a caret after the first nonzero digit, the 1. We then count the number of decimal places from the caret to the original decimal point. There are two decimal places, so we move the decimal point two places to the right to the caret and use 10^{-2}. Our answer is $0.01503 = 1.503 \times 10^{-2}$.

EXAMPLE

Change the following standard notation numbers to scientific notation:

a. 600,000,000,000

b. 2,450

c. −10,080,000

d. 0.79

e. −0.000005505

SOLUTION

a. $600{,}000{,}000{,}000 = 6 \times 10^{11}$.

b. $2{,}450 = 2.45 \times 10^3$.

c. $-10{,}080{,}000 = -1.008 \times 10^7$.

d. $0.79 = 7.9 \times 10^{-1}$.

e. $-0.000005505 = -5.505 \times 10^{-6}$.

Multiplication and Division of Numbers in Scientific Notation

To multiply two numbers in scientific notation, we multiply the numbers and add the exponents of the 10's. Thus, $(a \times 10^m)(b \times 10^n) = ab \times 10^{m+n}$. This method works whether the exponents m and n are positive or negative. Thus, $(4 \times 10^4)(2 \times 10^5) = 4(2) \times 10^{4+5} = 8 \times 10^9$. When we multiply a times b, it is possible that the product becomes greater than 10. If so, we change it to scientific notation as well.

EXAMPLE

Multiply the following.

a. $(3 \times 10^8)(1.4 \times 10^7)$

b. $(-7 \times 10^4)(1.1 \times 10^{-6})$

c. $(-1.2 \times 10^{-2})^2$

d. $(-3 \times 10^{10})(-5 \times 10^{-4})$

SOLUTION

a. $(3 \times 10^8)(1.4 \times 10^7) = 4.2 \times 10^{15}$.

b. $(-7 \times 10^4)(1.1 \times 10^{-6}) = -7.7 \times 10^{-2}$.

c. $(-1.2 \times 10^{-2})^2 = (-1.2 \times 10^{-2})(-1.2 \times 10^{-2}) = 1.44 \times 10^{-4}$.

d. $(-3 \times 10^{10})(-5 \times 10^{-4}) = 15 \times 10^6 = (1.5 \times 10^1) \times 10^6 = 1.5 \times 10^7$.

To divide two numbers in scientific notation, we divide the numbers and subtract the exponents of the 10's (numerator exponent minus denominator exponent). Thus, $\dfrac{a \times 10^m}{b \times 10^n} = \dfrac{a}{b} \times 10^{m-n}$. This method works whether the exponents m and n are positive or negative. Thus, $\dfrac{6 \times 10^9}{2 \times 10^4} = \dfrac{6}{2} \times 10^{9-4} = 3 \times 10^5$. When we divide a by b, it is possible that the quotient becomes less than 1. If so, we change it to scientific notation as well.

EXAMPLE

Find the following quotients.

a. $\dfrac{3.2 \times 10^{15}}{2 \times 10^7}$

b. $\dfrac{7.5 \times 10^4}{2.5 \times 10^{-2}}$

c. $\dfrac{1.21 \times 10^{-3}}{-1.1 \times 10^{-6}}$

d. $\dfrac{-2 \times 10^{-3}}{-8 \times 10^{-2}}$

SOLUTION

a. $\dfrac{3.2 \times 10^{15}}{2 \times 10^7} = 1.6 \times 10^{15-7} = 1.6 \times 10^8$.

b. $\dfrac{7.5 \times 10^4}{2.5 \times 10^{-2}} = 3 \times 10^{4-(-2)} = 3 \times 10^6$.

c. $\dfrac{1.21 \times 10^{-3}}{-1.1 \times 10^{-6}} = -1.1 \times 10^{-3-(-6)} = -1.1 \times 10^3$.

d. $\dfrac{-2 \times 10^{-3}}{-8 \times 10^{-2}} = 0.25 \times 10^{-3-(-2)} = (2.5 \times 10^{-1}) \times 10^{-1} = 2.5 \times 10^{-2}$.

CHAPTER 3

Algebraic Operations

CHAPTER 3

ALGEBRAIC OPERATIONS

An important part of understanding any topic is knowing the vocabulary, and this is especially true in mathematics. This chapter introduces the vocabulary of algebra.

OPERATIONS WITH ALGEBRAIC EXPRESSIONS

Algebra uses letters to represent numbers. For example, in the formula for the surface area of a sphere, $A = 4\pi r^2$, A stands for the surface area, r stands for the radius of the sphere, and π is the Greek letter *pi*, with a value of approximately 3.14.

The letters A and r can represent many numbers, so they are called **variables**. The number 4 and letter π each represents only one number, so they are called **constants**.

Terms are constants, variables, or a product or quotient of constants and variables. For example, 5, $4x$, $-2xy$, and $\dfrac{3x^2}{y}$ are all terms. The constant part of the term is called a **coefficient**, so the coefficient of $4x$ is 4. **Algebraic expressions** consist of one or more terms. They can contain mathematical symbols such as $+$, $(\ )$, or $\sqrt{\ }$. Examples of algebraic expressions include

$$-8 \qquad x - 6 \qquad |2a + b| \qquad \sqrt{3x + 1} \qquad x^2 - xy + y^2$$

Typically, we **substitute** numbers for the variables in an expression and then do the arithmetic to get a numerical answer. This is called **evaluating the expression**.

EXAMPLE

Evaluate $2x - y$ for $x = -\dfrac{1}{2}$ and $y = -5$.

SOLUTION

$$2x - y = 2\left(\frac{-1}{2}\right) - (-5) = -1 + 5 = 4.$$

EXAMPLE

Evaluate $|4a^2 + b - 1|$ for $a = 0$ and $b = \frac{1}{3}$.

SOLUTION

$$\left|4(0) + \frac{1}{3} - 1\right| = \left|\frac{-2}{3}\right| = \frac{2}{3}.$$

Equivalent expressions always have the same value for all replacements of variables. The expression $2x - 3$ is equivalent to $2x + (-3)$, no matter what value x has. Likewise, $\frac{4x}{8} = \frac{x}{2}$. Try it. Several properties, discussed next, are helpful in determining whether expressions are equivalent.

Commutative Properties

The **commutative property** allows us to change the order when adding or multiplying.

Addition: For any real numbers a and b, $a + b = b + a$. Thus, we can say $3 + 5 = 5 + 3$.

Multiplication: For any real numbers a and b, $ab = ba$. Thus, $2 \cdot 7 = 7 \cdot 2$, which can also be written as $2(7) = 7(2)$. Parentheses next to a number or variable indicate multiplication.

EXAMPLE

Use the commutative property of addition to write an equivalent expression for $8x + 5y$.

SOLUTION

$8x + 5y = 5y + 8x.$

CHAPTER 3: ALGEBRAIC OPERATIONS

EXAMPLE

Explain why subtraction generally is not commutative. For what values of a and b is subtraction commutative?

SOLUTION

For subtraction to be communtative, $a - b = b - a$, which generally is not true.

It is true when $a = b$.

Associative Properties

The **associative property** allows us to regroup the order when adding or multiplying.

Addition: For any real numbers a, b, and c, $(a + b) + c = a + (b + c)$. Thus, $(2 + 5) + 3 = 2 + (5 + 3)$. Remember to perform the operation inside of parentheses first.

Multiplication: For any real numbers a, b, and c, $(ab)c = a(bc)$. Thus, $(3 \cdot 2) \cdot 4 = 3 \cdot (2 \cdot 4)$.

EXAMPLE

Use the associative property of multiplication to write an equivalent expression for $2a \cdot (3b \cdot 4c)$.

SOLUTION

$2a \cdot (3b \cdot 4c) = (2a \cdot 3b) \cdot 4c$.

Identity Properties

When we add 0 to any real number, that number does not change. The number 0 is called the **additive identity**. When we multiply any number by 1, that number does not change. The number 1 is called the **multiplicative identity**.

Addition: For any real number a, $a + 0 = 0 + a = a$. Thus, $5 + 0 = 0 + 5 = 5$.

Multiplication: For any real number, $a(1) = 1(a) = a$. Thus, $4 \cdot 1 = 1 \cdot 4 = 4$.

Inverse Properties

When we add 5 and -5, we get 0. In this case, we call negative 5 the **additive inverse** of 5.

When we multiply 5 and $\frac{1}{5}$, we get 1. In this case, we call $\frac{1}{5}$ the **multiplicative inverse** of 5.

Addition: For any real number a, $a + (-a) = 0$. Thus, if we add any expression to its negative (additive inverse), we get 0, so $3 + (-3) = 0$, and $(x + 2) + [-(x + 2)] = x + 2 + (-x - 2) = 0$.

Multiplication: For any real number a, $a \neq 0$, $a\left(\frac{1}{a}\right) = 1$. (Remember that division by 0 is not defined, which is why $a \neq 0$.) Thus, if we multiply any expression by its reciprocal (multiplicative inverse), we get 1, so $4\left(\frac{1}{4}\right) = 1$.

EXAMPLE

What is the additive inverse of $4x - y$?

SOLUTION

The additive inverse is the negative of the expression, or $-(4x - y) = -4x + y$. To show that this is true, add the expression to its additive inverse to be sure the sum is zero. $4x - y + (-4x + y) = 0$.

EXAMPLE

What is the multiplicative inverse of $4x - y$?

SOLUTION

The multiplicative inverse is the reciprocal, or $\frac{1}{4x - y}$, and, to check, $(4x - y)\left(\frac{1}{4x - y}\right) = 1$.

Distributive Property

The **distributive property** tells us that when multiplying a number by a sum, we can add first and then multiply, or multiply first and then add. So for any real numbers a, b, and c, it is true that $a(b + c) = ab + ac$. Thus, by adding first and then multiplying, we have $3(1 + 5) = 3(6) = 18$. By multiplying first and then adding, we get $3(1) + 3(5) = 3 + 15 = 18$, the same answer.

EXAMPLE

Multiply $5(x + 8)$.

SOLUTION

$5(x + 8) = 5x + 40$.

EXAMPLE

Multiply $-2x(y - 4 + \pi)$.

SOLUTION

$-2x(y - 4 + \pi) = -2xy + 8x - 2\pi x$.

The reverse of using the distributive property is called **factoring**. To factor an expression, we essentially write the addition or subtraction as a multiplication. When we factor in this way, we try to find a number or variable that divides evenly into all the terms.

EXAMPLE

Factor $2x + 10$.

SOLUTION

Since 2 divides into both 2 and 10, we factor out the 2.

We then get $2x + 10 = 2(x + 5)$

Note that if we don't pick the *greatest* common factor, we must factor again to get the answer. Thus, to factor $4y - 16$, if we choose 2 as the common factor (an obvious choice since both terms are even), we get $4y - 16 = 2(2y - 8)$. But $2y - 8$ also has both terms with even coefficients, so now we have $4y - 16 = 2 \cdot 2(y - 4) = 4(y - 4)$. How much quicker it would have been if we had just chosen 4 as the common factor to begin with.

EXAMPLE

Factor $ab - ac + a$.

SOLUTION

$ab - ac + a = a(b - c + 1)$. Note that a appears in each term.

If the problem instead had asked us to factor $ab - ac + 2$, we would have said it wasn't factorable. Some expressions are not factorable.

Terms with exactly the same variables and exponents but not necessarily the same coefficients, such as $3x$ and $8x$, $5x^2$ and $-x^2$ and $\frac{2}{3}xy$ and $\frac{-1}{2}xy$, are called **like terms**. The terms $5x$ and $4x^2$ are not like terms since the x's do not have the same exponents. When we combine like terms by adding the coefficients, we actually are factoring and using the distributive property.

EXAMPLE

Combine: $4x + 5x$.

SOLUTION

$4x + 5x = x(4 + 5) = 9x$.

We can now use the distributive property to simplify more complicated expressions.

EXAMPLE

Simplify $8x - (3x - 1)$.

SOLUTION

$$
\begin{aligned}
8x - (3x - 1) &= 8x - 1(3x - 1) \\
&= 8x - 3x + 1 \\
&= 5x + 1.
\end{aligned}
$$

EXAMPLE

Simplify $6y - [4(2y + 1) - 5(y - 2) - 1]$.

SOLUTION

$6y - [4(2y + 1) - 5(y - 2) - 1]$
$= 6y - (8y + 4 - 5y + 10 - 1)$ by the distributive law.
$= 6y - (3y + 13)$ by combining like terms.
$= 6y - 3y - 13$ by the distributive law.
$= 3y - 13$.

OPERATIONS WITH EXPONENTS

In the exponential notation a^n, a is the **base** and n is the **exponent**.

Definition of Positive Whole Number Exponents

The exponential notation a^n, where n is an integer greater than 1, means $\underbrace{a \cdot a \cdot a \cdot \ldots \cdot a \cdot a}_{n \text{ factors}}$. In other words, the factor a appears n times in the multiplication. Thus, 4^3 is called 4 to the third power because it can be written as a product where all three factors are 4, or $4^3 = 4 \cdot 4 \cdot 4 = 64$.

For all a, by definition, $a^1 = a$ and $a^0 = 1$.

EXAMPLE

Find $(4x)^2$.

SOLUTION

$(4x)^2 = (4x)(4x) = 16x^2$.

EXAMPLE

Find $3(-2x)^5$.

SOLUTION

$3(-2x)(-2x)(-2x)(-2x)(-2x) = 3(-32x^5) = -96 x^5$.

Definition of Negative Exponents

For any nonzero real number b and integer n, $b^{-n} = \dfrac{1}{b^n}$. Negative exponents mean reciprocals. Note that negative exponents do not make expressions negative.

EXAMPLE

Find 4^{-2}.

SOLUTION

$$4^{-2} = \frac{1}{4^2} = \frac{1}{16}.$$

EXAMPLE

Find $(-2)^{-3} - (-8)^{-1}$.

SOLUTION

$$(-2)^{-3} - (-8)^{-1} = \frac{1}{(-2)^3} - \left(\frac{1}{-8}\right) = -\frac{1}{8} + \frac{1}{8} = 0.$$

EXAMPLE

Find $-3(-4m)^{-4}$.

SOLUTION

$$-3(-4m)^{-4} = \frac{-3}{(-4m)^4} = \frac{-3}{256m^4}.$$

EXAMPLE

Find $\dfrac{x^{-1}}{y^{-2}} + (-2)^0$.

SOLUTION

$$\frac{x^{-1}}{y^{-2}} + (-2)^0 = \frac{y^2}{x} + 1.$$

Note how a negative exponent moves a factor in the numerator to the denominator and a factor in the denominator to the numerator, and then changes the exponents to positive.

Multiplying and Dividing Expressions with Exponents

When we multiply expressions with the same base in exponential notation, we add the exponents: $x^2 \cdot x^3 = x^{(2+3)} = x^5$. If we write this out, we see that $x^2 \cdot x^3 = (x \cdot x)(x \cdot x \cdot x) = x \cdot x \cdot x \cdot x \cdot x = x^5$.

When we divide expressions with the same base in exponential notation, we subtract the exponents (top exponent minus bottom exponent): $\dfrac{b^6}{b^2} = b^{(6-2)} = b^4$. If we write this out, we see that

$$\frac{b^6}{b^2} = \frac{b \cdot b \cdot b \cdot b \cdot \cancel{b} \cdot \cancel{b}}{\cancel{b} \cdot \cancel{b}} = b^4.$$

So, for any nonzero real numbers a and integers m and n, $a^m \cdot a^n = a^{m+n}$ and $\dfrac{a^m}{a^n} = a^{m-n}$.

EXAMPLE

Find $4^6 \cdot 4^{-5}$.

SOLUTION

$4^6 \cdot 4^{-5} = 4^{(6-5)} = 4^1 = 4.$

EXAMPLE

Find $(8x^2)(-3x^{-5})$.

SOLUTION

$(8x^2)(-3x^{-5}) = 8(-3)x^{(2-5)} = -24x^{-3} = \dfrac{-24}{x^3}.$

EXAMPLE

Find $\dfrac{30x^5 y^{-2}}{-6x^2 y^3}$.

SOLUTION

$$\frac{30x^5 y^{-2}}{-6x^2 y^3} = -5x^{5-2} y^{(-2-3)} = -5x^3 y^{-5} = \frac{-5x^3}{y^5}.$$

Raising Expressions with Exponents to a Power

When we raise expressions with exponents to a power, we use the definition of powers to determine the answer. Hence, $(x^2)^3 = (x^2)(x^2)(x^2) = x^6$. This suggests the following rule: $(a^m)^n = a^{mn}$.

For any nonzero real numbers a and b and integers m, n, and p, $(a^m b^n)^p = a^{mp} b^{np}$. And finally, $\left(\dfrac{a^m}{b^n}\right)^p = \dfrac{a^{mp}}{b^{np}}$.

EXAMPLE
Find $(3^4)^4$.

SOLUTION
$(3^4)^4 = 3^{4 \cdot 4} = 3^{16}$.

In the following solutions, note how the negative exponents move the numerator to the denominator and vice versa, and the negative exponents change to positives.

EXAMPLE
Find $(2x^2)^{-5}$.

SOLUTION
$(2x^2)^{-5} = 2^{-5} x^{(2 \cdot -5)} = \dfrac{1}{32} x^{-10} = \dfrac{1}{32 x^{10}}$.

EXAMPLE
Find $\left(-\dfrac{1}{2} x^{-4} y z^{-1}\right)^{-5}$.

SOLUTION
$\left(-\dfrac{1}{2} x^{-4} y z^{-1}\right)^{-5} = \dfrac{(-1)^{-5}}{(2)^{-5}} x^{20} y^{-5} z^5 = \dfrac{2^5}{(-1)^5} \cdot \dfrac{x^{20} z^5}{y^5} = \dfrac{-32 x^{20} z^5}{y^5}$.

EXAMPLE

Find $\left(\dfrac{x^2}{y^{-3}}\right)^4$.

SOLUTION

$$\left(\dfrac{x^2}{y^{-3}}\right)^4 = \dfrac{x^8}{y^{-12}} = x^8 y^{12}.$$

EXAMPLE

Find $\left(\dfrac{5x^5 y^{-1}}{3z^{-3}}\right)^{-2}$.

SOLUTION

$$\left(\dfrac{5x^5 y^{-1}}{3z^{-3}}\right)^{-2} = \dfrac{5^{-2} x^{-10} y^2}{3^{-2} z^6} = \dfrac{9y^2}{25x^{10} z^6}.$$

Hint: To help remember whether the exponents are added, subtracted, or multiplied, use the familiar powers of 2, where each value is the one before it multiplied by 2:

2^0	2^1	2^2	2^3	2^4	2^5	2^6
1	2	4	8	16	32	64

Then substitute 2 for the variable in a simple expression. For example, if you can't remember whether you add or multiply the exponents in $x^7 \cdot x^4$, just use something like $2^3 \cdot 2^2$ to jog your memory. You know that $2^3 = 8$ and $2^2 = 4$, so $2^3 \cdot 2^2$ should equal $8 \cdot 4 = 32$; thus, $2^3 \cdot 2^2 = 2^5$, reminding you that the exponents are added. The same thing can be done as a reminder of what to do with the exponents for division ($\dfrac{2^5}{2^3} = \dfrac{32}{8} = 4 = 2^2$ so the exponents are subtracted) and for raising a power to a power (($2^3)^2 = 8^2 = 64 = 2^6$ so the exponents are multiplied).

Radicals and Fractional Exponents

A **radical** is an expression of the form $\sqrt[n]{b}$, which denotes the *n*th **root** of a value *b*. The integer *n* is the **index** of the radical, and the value *b* is called the **radicand**. The index is omitted if $n = 2$ (square root). So $x^{1/2} = \sqrt{x}$, and $x^{a/b} = \sqrt[b]{x^a} = \left(\sqrt[b]{x}\right)^a$. **Fractional exponents** do not create fractions; they create roots.

It is important to realize that $25^{1/2} = \sqrt{25} = 5$. The answer is not ± 5. Square roots are *always positive*. This can be confusing. The solution to $x^2 = 25$ is $x = \pm 5$ because $x^2 = 25$ is an equation. But the value of $\sqrt{25}$ is 5 and *not* -5.

The series of laws for fractional exponents and radicals come from the rules for multiplying and dividing expressions with exponents and for raising expressions with exponents to a power. They can be summarized as follows:

$$\left(\sqrt[n]{a}\right)^n = a^{n/n} = a^1 = a$$

$$\sqrt[n]{ab} = \sqrt[n]{a} \cdot \sqrt[n]{b}$$

$$\sqrt[n]{\frac{a}{b}} = \frac{\sqrt[n]{a}}{\sqrt[n]{b}}, \; b \neq 0$$

$$\left(\sqrt[n]{a^m}\right) = \left(\sqrt[n]{a}\right)^m$$

$$\sqrt[m]{\sqrt[n]{a}} = \sqrt[mn]{a}$$

When we work with fractional exponents, we can treat them just like fractions in algebraic expressions. So $\sqrt[4]{36^2} = 36^{2/4} = 36^{1/2} = \sqrt{36} = 6$.

EXAMPLE

Find $\left(36x^{10}\right)^{1/2}$.

SOLUTION

$\left(36x^{10}\right)^{1/2} = \sqrt{36}\,x^{10/2} = 6x^5$.

EXAMPLE

Find $(16x^{-2})^{3/4}$.

SOLUTION

$$(16x^{-2})^{3/4} = 16^{3/4} x^{-2(3/4)} = 16^{3/4} x^{-3/2} = \frac{(\sqrt[4]{16})^3}{x^{3/2}} = \frac{2^3}{x^{3/2}} = \frac{8}{x^{3/2}}.$$

EXAMPLE

Find $(27x^3)^{-2/3}$.

SOLUTION

$$(27x^3)^{-2/3} = \frac{1}{(27x^3)^{2/3}} = \frac{1}{(27^{2/3})\left(x^{3\left(\frac{2}{3}\right)}\right)} = \frac{1}{(\sqrt[3]{27})^2 x^2} = \frac{1}{3^2 x^2} = \frac{1}{9x^2}.$$

EXPANDING AND FACTORING POLYNOMIALS

A **polynomial** consists of one or more terms such that the terms are either constants or the product of constants and positive *integer* powers of variables. A polynomial consisting of only one term is called a **monomial**. A polynomial consisting of two terms is called a **binomial**. A polynomial consisting of three terms is called a **trinomial**.

$6x^3 + 4x - 3$ is a polynomial. Note that there is no x^2 term.

$9x + 4x^{1/2} - 1$ is not a polynomial since the second term's power is not an integer.

$8x^2 - 8x^{-2}$ is not a polynomial since the second term's power is not a positive integer.

The **degree** of a monomial is the sum of the exponents of the variables. The degree of a constant is zero.

$10x^3$ has degree 3.

$-2x^2yz^4$ has degree 7 ($= 2 + 1 + 4$).

12 has degree 0 because $x^0 = 1$, so the (implied) exponent of x is 0.

The degree of a polynomial is equivalent to the highest degree of any of its monomial terms.

$4x^3 + x^2 - 3x + 9$ has degree 3.

Multiplication of monomials is achieved by multiplying first the coefficients and then the variables by adding the powers, as we have already seen.

EXAMPLE

Multiply $(4x^2)(-3x^5)$.

SOLUTION

$(4x^2)(-3x^5) = -12x^7$.

EXAMPLE

Multiply $(-5a^3 bc^2)(-a^5c)$.

SOLUTION

$(-5a^3 bc^2)(-a^5c) = 5a^8bc^3$.

Multiplying a monomial by a polynomial uses the distributive property: the monomial gets multiplied by every term in the polynomial and, if possible, like terms are collected.

EXAMPLE

Multiply $x(4x - 5)$.

SOLUTION

$x(4x - 5) = 4x^2 - 5x$.

EXAMPLE

Multiply $-5x^2y(4x - 2xy + y)$.

SOLUTION

$-5x^2y(4x - 2xy + y) = -20x^3y + 10x^3y^2 - 5x^2y^2$.

FOIL Method

Multiplying two polynomials takes time. We have to multiply every term in the first polynomial by every term in the second polynomial and then collect like terms.

A special case involves multiplying two binomials. The method that is used is frequently called the **F-O-I-L** method, which stands for the products of the Firsts-Outers-Inners-Lasts.

The answer to $(x + 4)(2x - 5)$ has four terms:

Firsts: $x(2x) = 2x^2$
Outers: $x(-5) = -5x$
Inners: $4(2x) = 8x$
Lasts: $4(-5) = -20$.

So $(x + 4)(2x - 5) = 2x^2 - 5x + 8x - 20 = 2x^2 + 3x - 20$.

EXAMPLE
Multiply $(3x - 4)(4x - 1)$.

SOLUTION
$(3x - 4)(4x - 1) = 12x^2 - 3x - 16x + 4 = 12x^2 - 19x + 4$.

EXAMPLE
Find $(5x^2 - y)^2$.

SOLUTION
$(5x^2 - y)^2 = (5x^2 - y)(5x^2 - y)$
$= 25x^4 - 5x^2y - 5x^2y + y^2 = 25x^4 - 10x^2y + y^2$.

Several FOIL-type problems occur often enough that they deserve special recognition:

Sum and difference of same terms:	$(a + b)(a - b) = a^2 - b^2$.
Square of a binomial:	$(a + b)^2 = a^2 + 2ab + b^2$
	$(a - b)^2 = a^2 - 2ab + b^2$
Cube of a binomial:	$(a + b)^3 = a^3 + 3a^2b + 3ab^2 + b^3$
	$(a - b)^3 = a^3 - 3a^2b + 3ab^2 - b^3$
Sum and difference of cubes:	$a^3 + b^3 = (a + b)(a^2 - ab + b^2)$
	$a^3 - b^3 = (a - b)(a^2 + ab + b^2)$

EXAMPLE

Find a. $(x + 4)(x - 4)$
b. $(x - 4)^2$
c. $(x - 4)^3$

SOLUTION

a. $(x + 4)(x - 4) = x^2 - 16$.
b. $(x - 4)^2 = x^2 - 8x + 16$.
c. $(x - 4)^3 = x^3 - 12x^2 + 48x - 64$.

EXAMPLE

Find $(x - 4)(x^2 - 5x + 3)$.

SOLUTION

This isn't a FOIL problem because it doesn't involve two binomials.

Both terms in $(x - 4)$ must be multiplied by every term in the trinomial.

$(x - 4)(x^2 - 5x + 3) = x(x^2 - 5x + 3) - 4(x^2 - 5x + 3)$
$= x^3 - 5x^2 + 3x - 4x^2 + 20x - 12 = x^3 - 9x^2 + 23x - 12$.

Division of a Polynomial by a Polynomial

Division of a polynomial by a polynomial is a tedious procedure called long division, which is similar to long division of two integers.

Step 1: The terms of both polynomials are arranged in order of descending powers of one variable.

Step 2: The first term of the dividend is divided by the first term of the divisor, which gives the first term of the quotient.

Step 3: The divisor is multiplied by the first term of the quotient and the result is subtracted from the dividend.

Step 4: Using the remainder obtained from step 3 plus the next term of the original dividend as the new dividend, steps 2 and 3 are repeated until the remainder is zero or the degree of the remainder is less than the degree of the divisor.

Step 5: The result is written as follows:
$$\frac{\text{dividend}}{\text{divisor}} = \text{quotient} + \frac{\text{remainder}}{\text{divisor}}, \text{divisor} \neq 0.$$

Note: The remainder theorem, introduced in Chapter 4, will find the remainder (but not the quotient) without long division.

EXAMPLE

Find $(3x^2 - 4x + 2) \div (x - 2)$.

SOLUTION

$$\begin{array}{r} 3x+2 \\ x-2{\overline{\smash{\big)}\,3x^2-4x+2}} \\ \underline{-(3x^2-6x)} \\ 2x+2 \\ \underline{-(2x-4)} \\ 6 \end{array}$$

Thus, $(3x^2 - 4x + 2) \div (x - 2) = 3x + 2 + \dfrac{6}{x-2}$.

Factoring a Polynomial

The process of writing a polynomial as a product is called **factoring**. Unless noted otherwise, when factoring a polynomial, we want factors with *integer* coefficients. If a polynomial cannot be factored, it is considered to be **prime**.

To factor a trinomial in the form $ax^2 + bx + c$, we use the following pattern, which is a reverse of the FOIL method:

$$ax^2 + bx + c = (\square x + \square)(\square x + \square)$$

where the outer boxes are factors of a and the inner boxes are factors of c.

The goal is to find a combination of factors of a and c such that the outer and inner products add up to the middle term bx. According to the particular polynomial, determine which method below to take to make the process somewhat easier.

Method 1: Use this one when $a = 1$ and the sign of c is positive. This is the easiest case. The factors of the polynomial both have the same sign as the coefficient of b, their sum equals the coefficient of b, and their product equals c. For example, to find the factors of $x^2 + 7x + 10$, look at the factors of 10 (the value of c) to see which have a sum of 7 (the value of b). These are 5 and 2, so the factors are $(x + 2)(x + 5)$.

Method 2: Use this when $a = 1$ and the sign of c is negative. The factors of the polynomial have different signs. Again, their sum equals the coefficient of b, and their product equals c. For example, to find the factors of $x^2 - 3x - 4$, look at the factors of -4 (the value of c) to see which have a sum of -3 (the value of b). These are -4 and 1, so the factors are $(x - 4)(x + 1)$.

Method 3: Use this when $a \neq 1$. Write the whole number factors of a in one row and the integer number factors of c in another row, and check which combination adds to b. This can be very tedious because there can be many combinations to check. However, once we have found one that works, we don't have to continue checking the others. Following this method for $6x^2 + 7x - 3$, we get:

The factors of $a = 6$ are: 3 and 2, and 6 and 1.
The factors of $c = -3$ are: -3 and 1, and -1 and 3.

Even though there are only four pairs of factors, this yields eight combinations to add to see whether the sum is $+7$:

$3(-3) + 2(1)$ $3(1) + 2(-3)$ $3(-1) + 2(3)$ $3(3) + 2(-1)$

$6(-3) + 1(1)$ $6(1) + 1(-3)$ $6(-1) + 1(3)$ $6(3) + 1(-1)$

By the time we get to the last combination on the first row, however, we have the answer since $9 - 2 = 7$, and we don't have to go any further. So the factors of $6x^2 + 7x - 3$ are $(3x - 1)(2x + 3)$. Note the placement of the coefficients in the factors: they have to match the Inners and Outers of the FOIL method.

Method 4: Use the quadratic formula which will be introduced in Chapter 5.

Hint: If $a \neq 1$, and the CLEP College Algebra exam question is multiple choice, it is often easiest to multiply out the factors in each answer choice by using FOIL to see which is the correct one. Again, once we have found one that works, we don't have to continue checking the others. Imagine the time it takes to use step 3 to find the factors for an expression such as $12x^2 + 23x - 9$.

EXAMPLE

Factor $x^2 - 9x + 20$.

SOLUTION

Since $a = 1$, b is negative, and c is positive, we are looking for two negative numbers that are factors of 20 and add up to -9. These would be -4 and -5, and the answer is $(x - 4)(x - 5)$.

EXAMPLE

Factor $x^2 - 11x - 60$.

SOLUTION

Again, $a = 1$, but since c is negative, we want two numbers of opposite signs that multiply to -60 and have a sum of -11. These would be 4 and -15, so the factorization is $(x - 15)(x + 4)$.

EXAMPLE

Find the factors of $6x^2 + 5x - 4$.

SOLUTION

The factors of $a = 6$ are: 3 and 2, and 6 and 1.

The factors of $c = -4$ are: -4 and 1, -1 and 4, and 2 and -2.

Even though there are only five pairs of factors, this yields twelve combinations to check to see which combination gives $+5x$ as the middle term:

$$(3x - 4)(2x + 1) \qquad (3x + 4)(2x - 1)$$
$$(3x + 1)(2x - 4) \qquad (3x - 1)(2x + 4)$$
$$(3x + 2)(2x - 2) \qquad (3x - 2)(2x + 2)$$
$$(6x - 4)(x + 1) \qquad (6x + 4)(x - 1)$$

$$(6x + 1)(x - 4) \quad (6x - 1)(x + 4)$$
$$(6x + 2)(x - 2) \quad (6x - 2)(x + 2)$$

The correct combination (by trial and error) is
$6x^2 + 5x - 4 = (3x + 4)(2x - 1)$.

How tedious! But if the question is multiple choice, for a problem with so many possibilities, a good test-taking technique, again, is to use FOIL on the choices until you get one that multiplies to $6x^2 + 5x - 4$.

EXAMPLE

Factor $3x^2 - 4x - 15$.

SOLUTION

The factors of $a = 3$ are: 1 and 3.

The factors of $c = -15$ are: 1 and -15, -1 and 15, 3 and -5, -3 and 5.

This yields eight combinations to check to see which combination gives $-4x$ as the middle term.

$$(3x + 1)(x - 15) \quad (3x - 1)(x + 15)$$
$$(3x + 3)(x - 5) \quad (3x - 3)(x + 5)$$
$$(x + 1)(3x - 15) \quad (x - 1)(3x + 15)$$
$$(x + 3)(3x - 5) \quad (x - 3)(3x + 5)$$

The correct combination (by trial and error) is
$3x^2 - 4x - 15 = (x - 3)(3x + 5)$.

Again, it's so much easier to just look at the answer choices and pick the one that, by using the FOIL method, yields $3x^2 - 4x - 15$.

EXAMPLE

Factor $12x^2 + 25x + 12$.

SOLUTION

Again, for a problem with so many possibilities, a good test-taking technique is to look at each answer choice to see which multiplies to $12x^2 + 25x + 12$. The correct answer is $(3x + 4)(4x + 3)$.

Special Polynomial Factorizations

Factoring can be made easier by recognizing the special rules stated earlier in the discussion of the FOIL method and adapting them to factorization:

Grouping: $\quad a(c + d) + b(c + d) = (c + d)(a + b)$
$\quad\quad\quad\quad\quad\quad a(c + d) - b(c + d) = (c + d)(a - b)$

Common factor: $\quad ab + ac = a(b + c)$

Difference of squares: $\quad a^2 - b^2 = (a + b)(a - b)$.

Note that the sum of squares is prime.

Binomial squares: $\quad a^2 + 2ab + b^2 = (a + b)^2$
$\quad\quad\quad\quad\quad\quad a^2 - 2ab + b^2 = (a - b)^2$

Binomial cubes: $\quad a^3 + 3a^2b + 3ab^2 + b^3 = (a + b)^3$
$\quad\quad\quad\quad\quad\quad a^3 - 3a^2b + 3ab^2 - b^3 = (a - b)^3$

Sum and difference of cubes: $\quad (a + b)(a^2 - ab + b^2) = a^3 + b^3$
$\quad\quad\quad\quad\quad\quad (a - b)(a^2 + ab + b^2) = a^3 - b^3$

EXAMPLE
Factor $4x^2 - 25y^2$.

SOLUTION
This is a difference of squares, so $4x^2 - 25y^2 = (2x + 5y)(2x - 5y)$.

EXAMPLE
Factor $x^2 + 64$.

SOLUTION
$x^2 + 64$ is a sum of squares and is prime.

EXAMPLE
Factor $x^3 + 64$.

SOLUTION
$x^3 + 64$ is sum of cubes and is factorable: $x^3 + 64 = (x + 4)(x^2 - 4x + 16)$.

EXAMPLE

Completely factor $x^3 + 3x^2 - 4x - 12$.

SOLUTION

Trinomials can often be grouped. This one can be grouped as:

$x^3 + 3x^2 - 4x - 12 = (x^3 + 3x^2) - (4x + 12)$

Now, factoring out the common factor $(x + 3)$ from each group, we get:

$x^2(x + 3) - 4(x + 3) = (x + 3)(x^2 - 4)$.

Since $x^2 - 4$ is a difference of squares, we get:

$x^3 + 3x^2 - 4x - 12 = (x + 3)(x + 2)(x - 2)$.

EXAMPLE

Completely factor $12x^2 - 81x - 120$.

SOLUTION

To make this easier, we factor out the common factor 3.

$12x^2 - 81x - 120 = 3(4x^2 - 27x - 40)$.

For $4x^2$, the combination is either $2x$ and $2x$ or x and $4x$.

For -40, the combinations are 1 and -40, -1 and 40, 2 and -20, -2 and 20, 4 and -10, -4 and 10, 5 and -8, or -5 and 8.

By trial and error, the factorization is $12x^2 - 81x - 120 = 3(x - 8)(4x + 5)$.

Again, if the CLEP exam question is multiple choice, for a problem with so many possibilities, a good test-taking technique is to use FOIL on each answer choice to see which multiplies to $12x^2 - 81x - 120$.

OPERATIONS WITH RATIONAL EXPRESSIONS

A **rational expression** is an algebraic expression that can be written as the quotient of two polynomials: $\frac{p}{q}$, where $q \neq 0$. Examples of rational expressions are $\frac{3}{4}$, $3x + 5$, $\frac{x+2}{5}$, and $\frac{4x^2 + 5x - 1}{2x - 4}$ (where $x \neq 2$).

Multiplication of Rational Expressions

To multiply two rational expressions, we multiply numerators together to get the product numerator, and the denominators together to get the product denominator: $\frac{a}{b} \cdot \frac{c}{d} = \frac{ac}{bd}$.

Any factor in any numerator can cancel with the same factor in any denominator.

Note: Cancellation across rational expressions occurs only when the expressions are multiplied, *never* when they are added or subtracted.

EXAMPLE

Find $\frac{5x}{2y^2} \cdot \frac{-6y}{35x^3}$.

SOLUTION

$$\frac{5x}{2y^2} \cdot \frac{-6y}{35x^3} = \frac{-30xy}{70x^3y^2} = \frac{-3}{7x^2y}.$$

EXAMPLE

Find $\frac{x^2-9}{4x-12} \cdot \frac{-2x-2}{x^2+4x+3}$.

SOLUTION

$$\frac{x^2-9}{4x-12} \cdot \frac{-2x-2}{x^2+4x+3} = \frac{\cancel{(x+3)}\,\cancel{(x-3)}}{4\cancel{(x-3)}} \cdot \frac{-2\cancel{(x+1)}}{\cancel{(x+1)}\,\cancel{(x+3)}} = -\frac{2}{4} = -\frac{1}{2}$$

Division of Rational Expressions

To divide two rational expressions, we multiply by the reciprocal of the second expression. This is similar to dividing two fractions (invert-and-multiply rule): $\frac{a}{b} \div \frac{c}{d} = \frac{a}{b} \cdot \frac{d}{c} = \frac{ad}{bc}$.

EXAMPLE

Find $\dfrac{5x+20}{x^2-3x-4} \div \dfrac{x^3}{x^2+x}$.

SOLUTION

$$\dfrac{5x+20}{x^2-3x-4} \div \dfrac{x^3}{x^2+x} = \dfrac{5(x+4)}{(x-4)(x+1)} \cdot \dfrac{x(x+1)}{x^3} = \dfrac{5(x+4)}{x^2(x-4)}.$$

Addition and Subtraction of Rational Expressions

To add and subtract rational expressions, it is necessary that each rational expression has the same denominator. If they do, we can use the rule: $\dfrac{a}{c}+\dfrac{b}{c}=\dfrac{a+b}{c}$.

EXAMPLE

Add $\dfrac{3x}{5}+\dfrac{2y}{5}$.

SOLUTION

$$\dfrac{3x}{5}+\dfrac{2y}{5}=\dfrac{3x+2y}{5}.$$

EXAMPLE

Combine $\dfrac{x^2}{x^2-1}+\dfrac{6x}{x^2-1}-\dfrac{7}{x^2-1}$.

SOLUTION

$$\dfrac{x^2}{x^2-1}+\dfrac{6x}{x^2-1}-\dfrac{7}{x^2-1}=\dfrac{x^2+6x-7}{x^2-1}=\dfrac{(x+7)(x-1)}{(x+1)(x-1)}=\dfrac{x+7}{x+1}.$$

If the denominators of the rational expressions are not the same, however, it is necessary to find a **lowest common denominator (LCD)**. The LCD is the **least common multiple (LCM)** of the denominators. To find the LCM of two or more algebraic expressions, we first factor each expression. We then use each factor the greatest number of times it occurs in any of the factorizations.

To then add the rational expressions, we multiply each fraction by a ratio equal to 1 (e.g., $\frac{4}{4}$) so that each fraction has the same denominator (the LCD).

EXAMPLE

Find $\dfrac{x}{4}+\dfrac{x}{3}-\dfrac{x}{2}$.

SOLUTION

The LCD is 12.

So $\dfrac{x}{4}+\dfrac{x}{3}-\dfrac{x}{2} = \dfrac{x}{4}\left(\dfrac{3}{3}\right)+\dfrac{x}{3}\left(\dfrac{4}{4}\right)-\dfrac{x}{2}\left(\dfrac{6}{6}\right) = \dfrac{3x+4x-6x}{12} = \dfrac{x}{12}$.

EXAMPLE

Find $\dfrac{x-1}{3x+15}-\dfrac{x+2}{6x+30}$.

SOLUTION

$\dfrac{x-1}{3x+15}-\dfrac{x+2}{6x+30} = \dfrac{x-1}{3(x+5)}-\dfrac{x+2}{6(x+5)}$

The LCD is $6(x+5)$

So $\dfrac{x-1}{3(x+5)}\left(\dfrac{2}{2}\right)-\dfrac{x+2}{6(x+5)} = \dfrac{2x-2-x-2}{6(x+5)} = \dfrac{x-4}{6(x+5)}$.

EXAMPLE

Find $\dfrac{1}{2x+2}-\dfrac{x}{2x-4}+\dfrac{x^2+2}{x^2-x-2}$.

SOLUTION

$\dfrac{1}{2x+2}-\dfrac{x}{2x-4}+\dfrac{x^2+2}{x^2-x-2} = \dfrac{1}{2(x+1)}-\dfrac{x}{2(x-2)}+\dfrac{x^2+2}{(x+1)(x-2)}$

The LCD is $2(x+1)(x-2)$.

So $\dfrac{1}{2(x+1)}\left(\dfrac{x-2}{x-2}\right) - \dfrac{x}{2(x-2)}\left(\dfrac{x+1}{x+1}\right) + \dfrac{x^2+2}{(x+1)(x-2)}\left(\dfrac{2}{2}\right)$

$= \dfrac{x-2-x^2-x+2x^2+4}{2(x+1)(x-2)} = \dfrac{x^2+2}{2(x+1)(x-2)}$

Complex Fractions

Algebra sometimes uses **complex fractions**, which are fractions within fractions. Answers never contain complex fractions, so they must be eliminated. There are two methods to eliminate complex fractions:

Method 1: When the problem is in the form of $\dfrac{\frac{a}{b}}{\frac{c}{d}}$, we can rewrite this as $\dfrac{a}{b} \div \dfrac{c}{d}$, which, as we saw in the division of rational expressions, is the same as $\dfrac{a}{b} \cdot \dfrac{d}{c} = \dfrac{ad}{bc}$.

However, this does not work when the numerator and denominator are not single fractions.

Method 2: The best way to eliminate the complex fractions in *all* cases is to find the LCD of all the fractions in the complex fraction. Multiply all terms in the numerator and denominator by this LCD, and the fraction that is left is magically no longer complex.

EXAMPLE

Simplify $\dfrac{1+\frac{2}{3}}{1+\frac{5}{6}}$.

SOLUTION

The LCD is 6.

$\left(\dfrac{1+\frac{2}{3}}{1+\frac{5}{6}}\right)\left(\dfrac{6}{6}\right) = \dfrac{6\left(1+\frac{2}{3}\right)}{6\left(1+\frac{5}{6}\right)} = \dfrac{6+4}{6+5} = \dfrac{10}{11}$.

CHAPTER 3: ALGEBRAIC OPERATIONS

EXAMPLE

Simplify $\dfrac{\dfrac{1}{x}+\dfrac{1}{y}}{x+y}$.

SOLUTION

The LCD is xy.

$$\left(\dfrac{\dfrac{1}{x}+\dfrac{1}{y}}{x+y}\right)\left(\dfrac{xy}{xy}\right) = \dfrac{xy\left(\dfrac{1}{x}+\dfrac{1}{y}\right)}{xy(x+y)} = \dfrac{y+x}{xy(x+y)} = \dfrac{1}{xy}.$$

Eliminating Roots in the Denominator

Expressions with square roots in the denominator, such as $\dfrac{a}{\sqrt{b}}$, must be **rationalized**. This means that the square root must be removed from the denominator. To accomplish this, we multiply the expression by $\dfrac{\sqrt{b}}{\sqrt{b}}$, which equals 1 so it doesn't alter the expression.

EXAMPLE

Simplify $\dfrac{5}{2\sqrt{3}}$.

SOLUTION

$$\dfrac{5}{2\sqrt{3}}\left(\dfrac{\sqrt{3}}{\sqrt{3}}\right) = \dfrac{5\sqrt{3}}{2(3)} = \dfrac{5\sqrt{3}}{6}.$$

If the denominator has a sum or difference of expressions with radicals, to rationalize the expression we need to multiply numerator and denominator by the **conjugate** of the radical expression. The conjugate of $\sqrt{a}+\sqrt{b}$ is simply $\sqrt{a}-\sqrt{b}$, and the conjugate of $\sqrt{a}-\sqrt{b}$ is $\sqrt{a}+\sqrt{b}$. Note that the radicals in the denominator disappear because it is now the difference of two squares.

EXAMPLE

Simplify $\dfrac{6}{2+\sqrt{3}}$.

SOLUTION

The conjugate is $2-\sqrt{3}$.

$$\left(\dfrac{6}{2+\sqrt{3}}\right)\left(\dfrac{2-\sqrt{3}}{2-\sqrt{3}}\right) = \dfrac{6(2-\sqrt{3})}{2^2-(\sqrt{3})^2} = \dfrac{6(2-\sqrt{3})}{4-3} = 6(2-\sqrt{3}).$$

EXAMPLE

Simplify $\dfrac{\sqrt{5}+\sqrt{3}}{\sqrt{5}-\sqrt{3}}$.

SOLUTION

The conjugate is $\sqrt{5}+\sqrt{3}$.

$$\left(\dfrac{\sqrt{5}+\sqrt{3}}{\sqrt{5}-\sqrt{3}}\right)\left(\dfrac{\sqrt{5}+\sqrt{3}}{\sqrt{5}+\sqrt{3}}\right) = \dfrac{5+2\sqrt{15}+3}{5-3} = \dfrac{8+2\sqrt{15}}{2} = \dfrac{2(4+\sqrt{15})}{2} = 4+\sqrt{15}.$$

SOLVING LINEAR EQUATIONS AND INEQUALITIES

Linear Equations

A **linear equation** in one variable is one that can be put into the form $ax + b = 0$, where $a \neq 0$.

The general rule for solving an equation is that any operation performed on one side of an equation must be performed on the other side. With terms that are added to or subtracted from the equation, we can move a term to the opposite side of the equation as long as we change its sign. This is useful when trying to get a value for a variable by putting the variable alone on one side of the equation.

For multiplication and division, just perform the operations on both sides of the equation. There is no such "shortcut."

EXAMPLE

Solve for x: $4(x + 3) = 2(x - 5)$.

SOLUTION

$4x + 12 = 2x - 10$

$4x - 2x = -10 - 12$

$2x = -22$

$x = -11$.

EXAMPLE

Solve for x: $3x - 2(x + 4) = 4(x - 7) - 7$.

SOLUTION

$3x - 2x - 8 = 4x - 28 - 7$

$3x - 2x - 4x = -28 - 7 + 8$

$-3x = -27$

$x = 9$.

Equations Containing Fractions

If the equation contains fractions, it is best to clear them out by multiplying *every term* by the LCD. Be sure to do this to every term on *both* sides of the equation.

EXAMPLE

Solve for x: $\dfrac{2x-3}{2} + \dfrac{3x-5}{4} = \dfrac{6x-1}{10}$.

SOLUTION

The LCD is 20.

So $20\left(\dfrac{2x-3}{2}\right) + 20\left(\dfrac{3x-5}{4}\right) = 20\left(\dfrac{6x-1}{10}\right)$

$10(2x - 3) + 5(3x - 5) = 2(6x - 1)$

$20x - 30 + 15x - 25 = 12x - 2$

Combining like terms (changing sign when crossing the equal sign), we get

$23x = 53$

$x = \dfrac{53}{23}$.

A special case occurs if the equation is in the form $\frac{a}{b} = \frac{c}{d}$. We can then use **cross-multiplication** to get $ad = bc$. (This is really just looking at multiplying by the LCD in another way.)

EXAMPLE

Solve for x: $\frac{5x-2}{3} = \frac{6x+1}{5}$.

SOLUTION

$5(5x - 2) = 3(6x + 1)$

$25x - 10 = 18x + 3$

$7x = 13$

$x = \frac{13}{7}$.

Equations Containing Radicals

If there is a radical in the equation, we can put the radical by itself on one side of the equation and then square both sides of the equation to eliminate it. Remember to check the answer in the *original* equation since squaring both sides can introduce **extraneous solutions**, solutions that are incorrect. So we need to check our solutions in the original equation.

EXAMPLE

Solve for x: $\sqrt{5x-6} = 3$.

SOLUTION

$\left(\sqrt{5x-6}\right)^2 = 3^2$

$5x - 6 = 9$

$5x = 15$

$x = 3$

Check: $\sqrt{5(3)-6} = \sqrt{9} = 3$.

EXAMPLE

Solve for x: $\sqrt{2x+2} + 6 = 2$.

SOLUTION

$\sqrt{2x+2} = -4$

$\left(\sqrt{2x+2}\right)^2 = (-4)^2$

$2x + 2 = 16$

$2x = 14$

$x = 7$.

Check: $\sqrt{2(7)+2} + 6 = \sqrt{16} + 6 = 4 + 6 = 10 \neq 2$.

Therefore, this equation has no solution.

Inequalities

An **inequality** is a statement that the value of one quantity or expression is greater than or less than that of another. There are five inequality symbols:

> greater than
< less than
≥ greater than or equal to
≤ less than or equal to
≠ not equal to

When we have inequalities with variables, there are certain values of the variable that may make the inequality true and others that make it false. We can show solutions to inequalities by using a number line. An open dot (○) at a value means the inequality does not include that value, and a closed dot (•) means that it does includes that value.

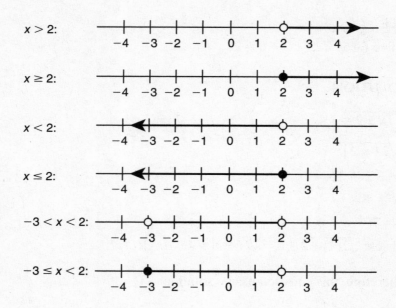

The rules for solving simple linear inequalities are the same as those for linear equations with one exception. If we multiply or divide both sides of an inequality by a negative number, the direction of the inequality changes. To check a solution, plug a number within the solution into the original inequality to see whether it is true. To be extra sure, you can choose a number *not* in the solution interval and be sure that it *does not* satisfy the original inequality.

EXAMPLE

Solve for x: $4x + 2 \geq 6x - 8$.

SOLUTION

Either of these solutions is correct.

$4x - 6x \geq -8 - 2$	$2 + 8 \geq 6x - 4x$
$-2x \geq -10$ or	$10 \geq 2x$
$x \leq 5$	$5 \geq x$

Check with $x = 0$: $4(0) + 2 \geq 6(0) - 8 \Rightarrow 2 \geq -8$, so the solution is correct.

Extra check with $x = 6$: $4(6) + 2 \geq 6(6) - 8 \Rightarrow 26 \geq 28$, but this is not true; since 6 is not in the solution set, our answer appears to be correct.

EXAMPLE

Solve for x: $\dfrac{3x}{4} - \dfrac{4x}{3} < 1$.

SOLUTION

$\dfrac{3x}{4} - \dfrac{4x}{3} < 1$

The LCD is 12.

$12\left(\dfrac{3x}{4}\right) - 12\left(\dfrac{4x}{3}\right) < 12(1)$

$9x - 16x < 12$

$-7x < 12$

$x > \dfrac{-12}{7}$.

Check with $x = 0$: $\dfrac{3(0)}{4} - \dfrac{4(0)}{3} < 1 \Rightarrow 0 < 1$, so the solution appears to be correct.

Extra check with $x = -12$: $\dfrac{3(-12)}{4} - \dfrac{4(-12)}{3} < 1 \Rightarrow -9 + 16 < 1 \Rightarrow 7 < 1$, which is not true; since -12 is not in the solution set, our answer appears to be correct.

EXAMPLE

Solve $5x - 9 \geq 3x + 3$.

SOLUTION

$5x - 3x \geq 9 + 3$

$2x \geq 12$

$x \geq 6$, or $[6, \infty)$ in interval notation (see Chapter 4)

Check with $x = 10$ (which is in the solution set): $5(10) - 9 \geq 3(10) + 3$ and $41 \geq 33$. True.

Extra check with $x = 0$ (which is not in the solution set): $5(0) - 9 \geq 3(0) + 3$, and since -9 is not greater than or equal to 3, this is further proof that the answer appears to be correct.

EXAMPLE

Solve $-1 - \dfrac{3x}{4} > x - 6$.

SOLUTION

First, multiply all terms by 4 to clear the fraction.

$-4 - 3x > 4x - 24$

$-4x - 3x > 4 - 24$

$-7x > -20$

$x < \dfrac{20}{7}$

(Note that dividing both sides of the inequality by a negative number (-7 here) changes the direction of the inequality sign.)

Check with $x = 0$, using the original inequality: $-1 > -6$, which is true.

We can solve **double inequalities** by solving each inequality separately.

EXAMPLE

Solve $-5 \leq 6x - 1 < 2$.

SOLUTION

$-5 \leq 6x - 1$ $6x - 1 < 2$

$-5 + 1 \leq 6x$ $6x < 1 + 2$

$-4 \leq 6x$ $6x < 3$

$\dfrac{-2}{3} \leq x$ $x < \dfrac{1}{2}$

Therefore, $-\dfrac{2}{3} \leq x < \dfrac{1}{2}$, or $\left[-\dfrac{2}{3}, \dfrac{1}{2}\right)$ in interval notation.

Any number between $-\dfrac{2}{3}$ inclusive and $\dfrac{1}{2}$ will make the inequality true.

Check with $x = 0$: $-5 \leq 6(0) - 1 < 2$ and $-5 \leq -1 < 2$, which is true.

CHAPTER 4

Functions and Their Properties

CHAPTER 4

FUNCTIONS AND THEIR PROPERTIES

RELATIONS AND GRAPHS

An **ordered pair** is commonly called a point. It is in the form of (x, y), where x and y are real numbers. A **relation** is a set of points.

EXAMPLE

For relation $R = \{(0, 0), (1, 0), (0, 1), (1, 1), (2, 0), (2, 1), (1, 2), (0, 2), (2, 2)\}$ list the set of ordered pairs for which the second member is greater than the first member.

SOLUTION

$\{(0, 1), (1, 2), (0, 2)\}$

The **domain** of a relation is the set of all first members of the relation. Even if a member is repeated, it is listed only once.

The **range** of a relation is the set of all second members of the relation. Again, even if a member is repeated, it is listed only once.

EXAMPLE

List the domain and range of the relation

$R = \{(-2, -2), (-1, 0), (-1, -2), (-1, 0)\}$

SOLUTION

Domain: $\{-2, -1\}$. Range: $\{-2, 0\}$.

Ordered pairs or points are illustrated on the Cartesian coordinate system. This is called a **graph**, and consists of the horizontal **x-axis** and the vertical **y-axis**, which intersect at the **origin**, the point (0,0). The axes divide the plane into four regions called **quadrants**. These are usually numbered by Roman numerals, counterclockwise from the upper right quadrant.

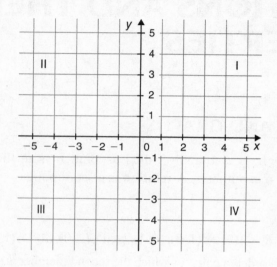

EXAMPLE

Graph the points $A(2, 4)$, $B(1, 0)$, $C(-3, -1)$, $D(-4, 2.5)$, and $E(1.5, -3)$

SOLUTION

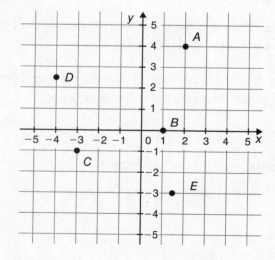

We can graph equations by plotting points and "connecting the dots." We usually plot at least five points to establish a pattern. Later in this chapter, we recognize common graphs by their equations.

EXAMPLE

Graph $y = 2x - 3$.

SOLUTION

Find some ordered pairs that satisfy the equation. Choose any number for x and calculate y. It is best to choose whole number values of x close to zero.

SOLUTION

x	y
0	−3
1	−1
2	1
−1	−5
−2	−7

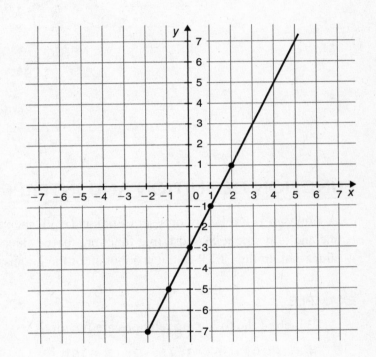

EXAMPLE

Graph $x = y^2 - 1$.

SOLUTION

Again, find some ordered pairs that satisfy the equation. In this case, it is easier to choose any number for y and calculate x. Remember, though, that points are still graphed in the order (x, y).

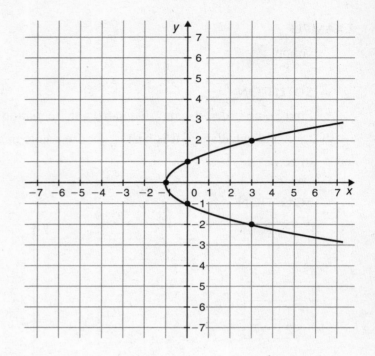

FUNCTION DEFINITION AND INTERPRETATION

A **function** is a relation, or a set of points (x, y), such that for every x, there is one and only one y. In short, in a function, the x-values cannot repeat but the y-values can. In the CLEP test, just about all of the graphs are of functions.

EXAMPLE

Which of the following relations are functions?

$A = \{(8, 0), (-6, -3), (1, -2), (2, -10)\}$

$B = \{(5, 1), (7, f), (5, b), (-3, p)\}$

$C = \{(1, 5), (-1, -7), (0, 5), (1, -1), (-1, -2)\}$

$D = \{(3, 4), (5, 4), (8, 4), (1, 4), (12, 4), (0, 4)\}$

SOLUTION

A and D are functions. No x- or y-values repeat in A, and even though $y = 4$ repeats in D, no x-values repeat. Relation B is not a function because $x = 5$ repeats. Relation C is not a function because both $x = 1$ and $x = -1$ repeat.

For any graph, we can quickly determine whether it is a function by the **vertical line test**: If it is possible for a vertical line to intersect a graph at more than one point, then the graph is not a function.

EXAMPLE

Use the vertical line test to determine which of the following graphs are functions.

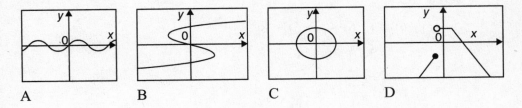

A   B   C   D

SOLUTION

Graph A is a function because any vertical line intersects the graph in only one location. Graphs B and C are not functions because it is possible to draw a vertical line that will intersect the graph in more than one location. Graph D is a function. An open dot indicates that the graph does not contain that point, and a closed dot indicates that the graph does include that point, so a vertical line at the value of x through the two dots still intersects the graph in only one location.

Functions can be represented in four different ways: numerical, graphical, symbolic, and verbal. In the **numerical representation**, functions are given by the set of points. The set $F = \{(-1, 1), (0, 2), (1, 3), (2, 4), (3, 5)\}$ is an example of a function given numerically.

A **graphical representation** gives a picture of the function. In the example above, A and D are cases of functions shown graphically.

The most usual way to represent a function is **symbolically**. For this representation, we use a special **function notation**. This notation is in the form of either "$y =$" or "$f(x) =$". In the $f(x)$ notation, we are stating a rule to find y given a value of x. Using this notation, it is easy to **evaluate the function** by plugging in a value of x to find y.

We discuss the **verbal** representation of functions in a separate section later in this chapter.

EXAMPLE

If $f(x) = x^2 - 5x + 8$, find

a. $f(-6)$

b. $f\left(\dfrac{3}{2}\right)$

c. $f(a)$

SOLUTION

a. $f(-6) = (-6)^2 - 5(-6) + 8$

$\quad = 36 + 30 + 8$

$\quad = 74.$

b. $f\left(\dfrac{3}{2}\right) = \left(\dfrac{3}{2}\right)^2 - 5\left(\dfrac{3}{2}\right) + 8$

$\quad = \dfrac{9}{4} - \dfrac{15}{2} + 8$

$\quad = \dfrac{11}{4}.$

c. $f(a) = a^2 - 5a + 8.$

Functions do not always use the variable x. In the CLEP test, other variables are used liberally.

EXAMPLE

If $A(r) = \pi r^2$, find

a. $A(3)$

b. $A(2s)$

c. $A(r + 1) - A(r)$

SOLUTION

a. $A(3) = 9\pi.$

b. $A(2s) = \pi(2s)^2 = 4\pi s^2.$

c. $A(r + 1) - A(r) = \pi(r + 1)^2 - \pi r^2$

$\quad = \pi(r^2 + 2r + 1 - r^2)$

$\quad = \pi(2r + 1).$

Composition of Functions

One concept that comes up in the CLEP exam is the **composition of functions**, which is a function of a function. So if we have two functions, f and g, we can find $f(g(a))$ or $g(f(a))$. These are different from $f(a) \cdot g(a)$, which is just the product of the functions. To find a composition of functions, plug a value into one function, determine the answer, and plug that answer into the second function.

EXAMPLE

If $f(x) = x^2 - x + 1$ and $g(x) = 2x - 1$, find

a. $f(-1) \cdot g(-1)$

b. $f(g(-1))$

c. $g(f(-1))$

SOLUTION

a. $f(-1) \cdot g(-1) = (1+1+1) \cdot (-2-1) = 3(-3) = -9$.

b. $g(-1) = 2(-1) - 1 = -3$, so $f(g(-1)) = f(-3) = 9 + 3 + 1 = 13$.

c. $f(-1) = 1 + 1 + 1 = 3$, so $g(f(-1)) = g(3) = 6 - 1 = 5$.

EXAMPLE

The graph of $g(x)$ is given in the figure below. Find the value of $g(g(-1))$.

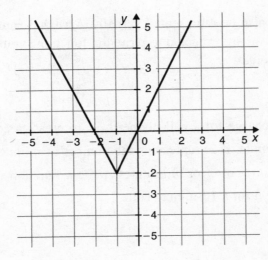

SOLUTION

$g(-1) = -2; g(g(-1)) = g(-2) = 0$.

Piecewise Functions

A **piecewise function** presents different rules based on the value of x.

EXAMPLE

If $f(x) = \begin{cases} x^2 - 3, & x \geq 0 \\ 2x + 1, & x < 0 \end{cases}$, find

a. $f(4)$

b. $f(3) - f(-1)$

c. $f(f(0))$

SOLUTION

a. For $f(4)$, since $4 \geq 0$, we use the function $x^2 - 3$, so $f(4) = 16 - 3 = 13$.

b. For $f(3)$, we use $x^2 - 3$ as the function, and since $-1 < 0$, for $f(-1)$ we use $2x + 1$ as the function. Therefore, $f(3) - f(-1) = (9-3) - (-2+1) = 6 - (-1) = 7$.

c. For $f(0)$, we use $x^2 - 3$ as the function $f(0) = 0 - 3 = -3$. Then for $f(-3)$ we use $2x + 1$. Therefore, $f(f(0)) = f(-3) = -6 + 1 = -5$.

Verbal Representation of a Function

A function can also be represented verbally. A **verbal representation** gives words from which we are expected to write the function symbolically and evaluate it at specific values.

EXAMPLE

A roaming charge for a cell phone is $1 to create a connection and 25 cents per minute. If m represents the number of minutes of a call and C represents the charge for m minutes, write a function that represents C, and evaluate it for a 10-minute call.

SOLUTION

$C(m) = 1 + 0.25m$,

$C(10) = 1 + 0.25(10) = 1 + 2.50 = 3.50$.

EXAMPLE

Considering rent, lighting, and paying employees, it costs a movie theater $1,400 for a showing of a movie. Let s be the number of seats sold for a showing of a movie with all seats costing $8. Let P represent the profit (or loss) for the theater selling s seats. Write a function that represents P and evaluate it for 225 people in the theater and for 75 people in the theater.

SOLUTION

The function is $P(s) = 8s - 1400$.

For 225 attendees, this would be $P(225) = 8(225) - 1400 = \$400$ (a profit).

For 75 attendees, this would be $P(75) = 8(75) - 1400 = -\$800$ (a loss).

Interval Notation

First, CLEP questions may ask about the behavior of functions in intervals, where intervals can be written with a description in terms of $<, \leq, >, \geq$ or $\neq$, or by using **interval notation**, as shown in the following table. Parentheses indicate "open" intervals, meaning the endpoint *is not* included. Brackets indicate "closed" intervals, meaning the endpoint *is* included. Note that an interval can have an open end and a closed end.

Also note that ∞ and $-\infty$ are always open-ended since we can never "reach" infinity. Therefore, any notation with a bracket rather than a parenthesis at the infinity side is incorrect.

Description	Interval notation	Description	Interval notation	Description	Interval notation
$x > a$	(a, ∞)	$x \leq a$	$(-\infty, a]$	$a \leq x < b$	$[a, b)$
$x \geq a$	$[a, \infty)$	$a < x < b$	(a, b) (open interval)	$a < x \leq b$	$(a, b]$
$x < a$	$(-\infty, a)$	$a \leq x \leq b$	$[a, b]$ (closed interval)	All real numbers	$(-\infty, \infty)$

If a solution is in one interval *or* the other, interval notation will use the connector ∪. So $x \leq 1$ or $x > 4$ would be written $(-\infty, 1] \cup (4, \infty)$ in interval notation. Solutions in intervals are usually written in the easiest way to define them. For instance, saying that $x < 0$ or $x > 0$ or $(-\infty, 0) \cup (0, \infty)$ is best expressed as just $x \neq 0$.

Domain

The **domain of a function** is the set of allowable x-values, as described earlier. The domain of a function f is $(-\infty, \infty)$ except for values of x that create a zero in the denominator, an even root of a negative number, or a logarithm of a nonpositive number. The domain of $a^{p(x)}$, where a is a positive constant and $p(x)$ is a polynomial, is $(-\infty, \infty)$.

EXAMPLE

Find the domain of the following functions:

a. $f(x) = x^2 - 3x - 1$

b. $f(x) = \dfrac{4}{x+2}$

c. $f(x) = \dfrac{2x}{x^2 + x - 20}$

d. $f(x) = \sqrt{x-8}$

SOLUTION

a. Since there is no denominator nor a radical, there are no restrictions and the domain is $(-\infty, \infty)$.

b. A function is undefined when the denominator equals zero, so $x + 2 \neq 0$. The domain $x \neq -2$ excludes this point.

CHAPTER 4: FUNCTIONS AND THEIR PROPERTIES

c. A function is undefined when the denominator equals 0, so $x^2 + x - 20 = (x + 5)(x - 4) \neq 0$.

The domain excludes both points, or $x \neq -5, x \neq 4$. In interval notation, this would be $(-\infty, -5) \cup (-5, 4) \cup (4, \infty)$.

d. A function is defined only when a radicand ≥ 0, so $x - 8 \geq 0$ and $x \geq 8$.

Range

The **range of a function** is the set of allowable y-values, as described earlier. Often, the range can be determined visually as the [lowest possible y-value, highest possible y-value]. Finding the range of some functions is fairly simple, such as the range of $y = x^2$; since any positive number squared is positive, the range is $[0, \infty)$. Likewise, the range of $y = \sqrt{x}$ is also all positives $[0, \infty)$ because the domain is $[0, \infty)$ and the square root of any positive number is positive. The range of $y = a^x$, where a is a positive constant, is $(0, \infty)$ since positive constants to powers must be positive (but $y \neq 0$ since $a \neq 0$).

EXAMPLE
Find the range of $f(x) = 4 - x^2$.

SOLUTION
Since the smallest value of x^2 is zero, the range of the function is $(-\infty, 4]$.

EXAMPLE
Find the range of $f(x) = \sqrt{x+1} + 1$.

SOLUTION
Since the smallest value of $\sqrt{x+1}$ is zero, the range of the function is $[1, \infty)$.

LINEAR FUNCTIONS

The types of functions that appear most frequently on the CLEP exam are **linear**, or **first-degree**, **functions**. The graph of any linear equation is a straight line.

Slope

The **slope** of a line, which is the ratio of the change in y to the change in x, measures the steepness of the line. It is commonly referred to as "rise over run."

Given two points (x_1, y_1) and (x_2, y_2), the slope of the line passing through the points can be written as:

$$m = \frac{\text{rise}}{\text{run}} = \frac{y_2 - y_1}{x_2 - x_1}.$$

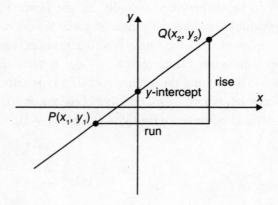

Lines that go up to the right (such as the line in the graph above) have a positive slope, whereas lines that go up to the left have a negative slope.

EXAMPLE

Find the slope of the lines passing though the following two points:

a. (2, 3) and (4, 7)

b. (−5, 2) and (1, −3)

SOLUTION

a. $m = \dfrac{7-3}{4-2} = \dfrac{4}{2} = 2.$

b. $m = \dfrac{-3-(2)}{1-(-5)} = \dfrac{-3-2}{1+5} = \dfrac{-5}{6}.$

Horizontal and Vertical Lines

Horizontal lines: Lines that are horizontal are in the form: $y = $ constant. Horizontal lines have zero slope. In the formula for slope, $y_2 - y_1 = 0$, so the slope is always zero.

EXAMPLE
Find the slope of the line passing through $(4, -8)$ and $(-2, -8)$.

SOLUTION
$$m = \frac{-8-(-8)}{-2-4} = \frac{-8+8}{-2-4} = \frac{0}{-6} = 0.$$ This is a horizontal line.

Vertical lines: Lines that are vertical are in the form: $x = $ constant. Vertical lines have an undefined slope; they are so steep that we cannot give a value to the steepness. In the formula for slope, $x_2 - x_1 = 0$, but division by zero is impossible.

EXAMPLE
Find the equations of the two lines in the figure below.

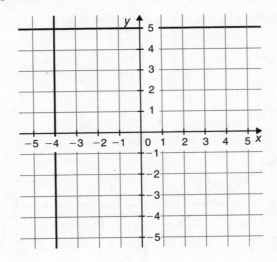

SOLUTION
The horizontal line is $y = 5$. The vertical line is $x = -4$.

Linear Equations

The equations of lines can have several forms, each having its own particular use.

General Form

The **general form** of an equation is $Ax + By + C = 0$, where A, B, and C are integers. This form of a linear equation gives the least amount of information about the line, but it is useful because it contains no fractions. An equation in this form often has to be put into the other forms, described below, to get more information about the line, such as its slope. Sometimes answers in the CLEP exam are given in general form.

Slope-Intercept Form

All lines except vertical lines will have a **y-intercept**—the point, in the form of $(0, b)$, at which the line crosses the y-axis. The equation of a line with slope m and y-intercept b is given by $y = mx + b$. If we are given the slope m and the y-intercept b, it is easy to write the equation of a line.

EXAMPLE

Find the equation of the line in the graph below.

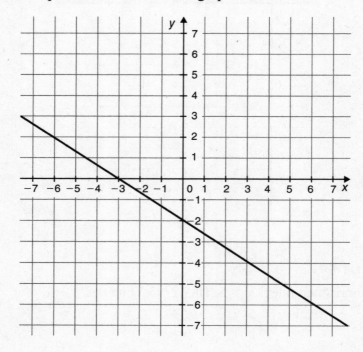

SOLUTION

We see that the y-intercept is $b = -2$. The line goes down 2 units for every 3 units it goes to the right so the rise is -2 for a run of 3. Therefore, the slope is $m = \frac{-2}{3}$. So the equation is $y = \frac{-2}{3}x - 2$. To get the general form, clear this equation of fractions by multiplying each term by 3 to get $3y = -2x - 6$, or $2x + 3y + 6 = 0$.

Point-Slope Form

The equation of a line with slope m and passing through the point (x_1, y_1) is given by $y - y_1 = m(x - x_1)$.

If we are given two points, this form is the easiest way to find the equation of the line. We need to find the slope first, then plug either point into the point-slope formula.

EXAMPLE

Find the equation passing through the points $(3, -2)$ and $(-1, 4)$.

SOLUTION

The slope is $m = \frac{4+2}{-1-3} = \frac{6}{-4} = -\frac{3}{2}$. Choosing $(-1, 4)$ as the point, the point-slope form gives $y - 4 = -\frac{3}{2}(x+1)$, or (multiplying through by 2), $2y - 8 = -3x - 3$. In general form, this is $3x + 2y - 5 = 0$.

EXAMPLE

Find the equation passing through the points $\left(\frac{1}{8}, -\frac{1}{2}\right)$ and $\left(\frac{3}{4}, 2\right)$.

SOLUTION

The slope $m = \left(\dfrac{2 + \frac{1}{2}}{\frac{3}{4} - \frac{1}{8}}\right)$. Multiply the numerator and the denominator by the LCD of 8 to clear the fractions. Then $m = \frac{16+4}{6-1} = \frac{20}{5} = 4$.

Choosing $\left(\frac{3}{4}, 2\right)$ as the point, $y - 2 = 4\left(x - \frac{3}{4}\right)$, or $y - 2 = 4x - 3$, which gives $y = 4x - 1$. The general form is thus $4x - y - 1 = 0$.

Intercept Form

In intercept form, the equation of a line with x-intercept a and y-intercept b is given by $\frac{x}{a} + \frac{y}{b} = 1$. If we know these intercepts, we can immediately write the equation of the line.

EXAMPLE

Find the equation passing through the points $(0, 4)$ and $(-3, 0)$.

SOLUTION

We can use the intercept form above, $\frac{x}{a} + \frac{y}{b} = 1$, where point $(0, 4)$ says the y-intercept is 4 and point $(-3, 0)$ says the x-intercept is -3. Since we know the intercepts, we can write the equation directly: $\frac{x}{-3} + \frac{y}{4} = 1$. Clearing this equation of fractions by multiplying through by the LCD $= 12$ gives $-4x + 3y = 12$, which in general form is $4x - 3y + 12 = 0$.

Parallel and Perpendicular Lines

Parallel lines are two distinct lines that have the same slope: $m_1 = m_2$. **Perpendicular lines** are two lines with slopes that are negative reciprocals: $m_2 = \frac{-1}{m_1}$, or $m_1 m_2 = -1$.

EXAMPLE

In the following graph, the solid line L is given by $x - 2y = 6$. Find the equations of the lines passing through the point $(2, 3)$ that are (a) parallel to L and (b) perpendicular to L.

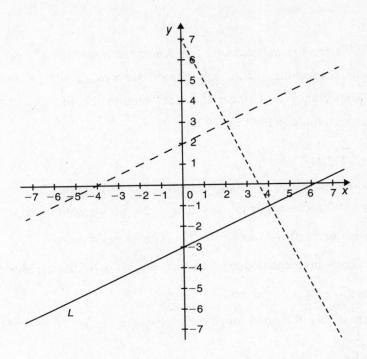

SOLUTION

Find the slope of line L by putting its equation (algebraically) into slope-intercept form: $x - 2y = 6 \Rightarrow 2y = x - 6 \Rightarrow y = \frac{1}{2}x - 3$, so $m = \frac{1}{2}$.

a. The parallel line has the same slope, $m = \frac{1}{2}$, and passes through the point (2, 3). Using the point-slope form,
$y - 3 = \frac{1}{2}(x - 2)$, or $y = \frac{1}{2}x + 2$.

b. The perpendicular line has a negative reciprocal slope, $m = -2$, so $y - 3 = -2(x - 2)$, or $y = -2x + 7$.

Representations of Verbal Problems

Linear equation representations of verbal problems usually involve finding several points that describe the situation, finding the equation of the line, and answering a question about the value of the function at another x-value.

EXAMPLE

In 1970, the life expectancy of an American male was 67 years. In 1990, the life expectancy of an American male was 72 years. Assuming that life expectancy is a linear function, predict the life expectancy of an American male in the year 2020.

SOLUTION

It is convenient to say that the year 1970 corresponds to $x = 0$ and the year 1990 corresponds to the year $x = 20$. So we want the equation of the line between $(0, 67)$ and $(20, 72)$. The slope is $m = \dfrac{72-67}{20-0} = \dfrac{5}{20} = \dfrac{1}{4}$, and the y-intercept (value of y when $x = 0$) is 67, so the slope-intercept form gives us $y = \dfrac{1}{4}x + 67$.

Alternatively, the point-slope form gives the line $y - 67 = \dfrac{1}{4}(x - 0)$, or $y = \dfrac{1}{4}x + 67$.

The year 2020 corresponds to $x = 50$ (that is, 50 years after 1970), so $y = \dfrac{1}{4}(50) + 67 = 79.5$.

EXAMPLE

A diver is 250 feet below the surface of the ocean when he starts to ascend. Two and a half minutes later, he is at a depth of 200 feet. Assuming that he ascends at a constant rate, predict the depth of the diver at eight and a half minutes after he starts to ascend.

SOLUTION

First we must decide whether to work in minutes or seconds. Minutes will be small numbers but involve fractions, whereas seconds will be large numbers but they are whole numbers, so let's use seconds. Two and a half minutes equals 150 seconds.

The text gives us the following points: $(0, -250)$, $(150, -200)$, so the slope is $m = \dfrac{-200+250}{150-0} = \dfrac{50}{150} = \dfrac{1}{3}$.

Again, we already have the intercept of -250, so it is easier to use the slope-intercept form of the equation: $y = \frac{1}{3}x - 250$.

Then, at 8.5 minutes (510 seconds), $y = \frac{1}{3}(510) - 250 = -80$. So the diver will then be 80 feet below the surface of the ocean.

Linear Inequalities

Linear inequalities are graphed by shading a section of the coordinate plane. The line is graphed using a solid line if the line is included ($\leq$ or $\geq$) or a dashed line if the line is not included ($<$ or $>$). Shading will be on either side of the line, and the solution is every point on the shaded side. If the inequality is in the form $y > f(x)$ or $y \geq f(x)$, the shading is above the line, and if the inequality is in the form $y < f(x)$ or $y \leq f(x)$, the shading is below the line.

EXAMPLE
Graph $y \geq 3x - 1$.

SOLUTION
We graph the line $y = 3x - 1$ and make it solid. Since the inequality is $\geq$, we shade above this line.

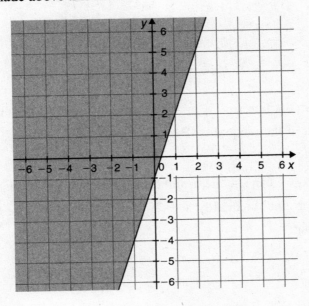

This means every point in the shaded region is true for $y \geq 3x - 1$. Let's check one random point, such as $(-3, 1)$, to show this is correct. We get $1 \geq 3(-3) - 1 = -10$. Is $1 \geq -10$? Yes, so it checks out.

EXAMPLE

Graph $x + 2y < 4$.

SOLUTION

$2y < -x + 4$
$y < -\frac{1}{2}x + 2$

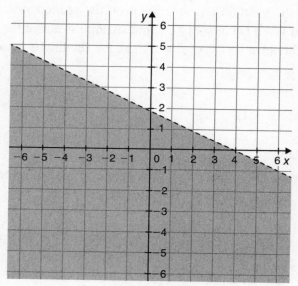

We graph the line $x + 2y = 4$ and make it a dashed line because it is not included in the inequality. Because the inequality for y is < 0 we shade below the line.

To check, we can use the origin $(0, 0)$ since it is in the shaded region. We use the original inequality, $x + 2y < 4$. Is $0 + 2(0) < 4$? Since $0 < 4$, it checks out.

GRAPHS OF COMMON FUNCTIONS

Certain graphs occur so often in algebra that students should know the general shape, where they hit the x-axis (zeros) and y-axis (y-intercept), and the domain and range. These graphs are commonly called **curves**, even though the graph may consist of straight lines.

Symmetry

Symmetry refers to the case in which a curve has the same exact shape on either side of a particular line. Not all graphs are symmetric.

A graph has **y-axis symmetry** if the curve is the same on either side of the y-axis. That means anything in the first quadrant would also appear in the second quadrant and vice versa. Also, anything in the third quadrant would also appear in the fourth quadrant and vice versa. Functions that are symmetric to the y-axis, such as the three following graphs, are said to be **even functions**.

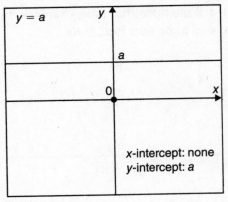

Domain: $(-\infty, \infty)$
Range: $[a, a]$
Symmetry: y-axis (even)

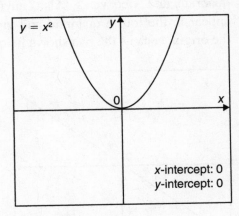

Domain: $(-\infty, \infty)$
Range: $[0, \infty)$
Symmetry: y-axis (even)

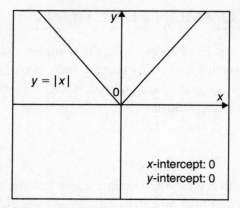

Domain: $(-\infty, \infty)$
Range: $[0, \infty)$
Symmetry: y-axis (even)

A graph has **origin symmetry** if the curve is the same on either side of the origin. That means anything in the first quadrant would also appear in the third quadrant and vice versa. Also, anything in the second quadrant would also appear in the fourth quadrant and vice versa. Functions that are symmetric to the origin, such as the two shown below, are said to be **odd functions.**

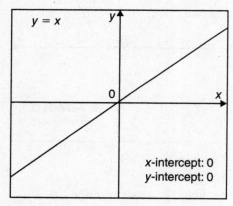

Domain: $(-\infty, \infty)$
Range: $(-\infty, \infty)$
Symmetry: origin (odd)

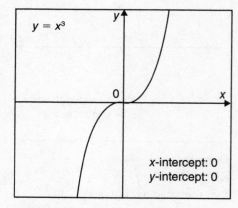

Domain: $(-\infty, \infty)$
Range: $(-\infty, \infty)$
Symmetry: origin (odd)

Graphs also can have **x-axis symmetry**, but these graphs would not be functions because they would fail the vertical line test.

Of course, many graphs have no symmetry, such as the one shown below.

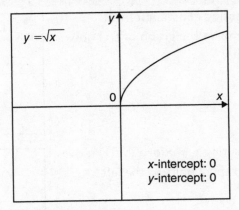

Domain: [0, ∞)
Range: (0, ∞)
Symmetry: none

TRANSFORMATION OF GRAPHS

A curve in the form $y = f(x)$, which is a basic common function, can be transformed in a variety of ways. The shape of the resulting curve stays the same, but the x- and y-intercepts might change, or the graph could be reversed. The table below describes transformations to a general function $y = f(x)$ with $f(x) = x^2$ as an example.

Notation	How $f(x)$ changes	Example with $f(x) = x^2$
$f(x) + a$	Adding constant a to $f(x)$ moves the graph of $f(x)$ up a units.	

Notation	How $f(x)$ changes	Example with $f(x) = x^2$
$f(x) - a$	Subtracting constant a from $f(x)$ moves the graph of $f(x)$ down a units.	
$f(x + a)$	Substituting $x + a$ for x in $f(x)$ moves the graph a units to the left.	
$f(x - a)$	Substituting $x - a$ for x in $f(x)$ moves the graph a units to the right.	
$a \cdot f(x)$	Multiplying $f(x)$ by $a > 1$ results in a vertical stretch.	
$a \cdot f(x)$	Multiplying $f(x)$ by a positive fraction, $0 < a < 1$, results in a vertical shrink.	

Notation	How $f(x)$ changes	Example with $f(x) = x^2$
$f(ax)$	Substituting ax for x in $f(x)$, where $a > 1$, results in a horizontal compression. (For this curve, the effect is the same as a vertical stretch.)	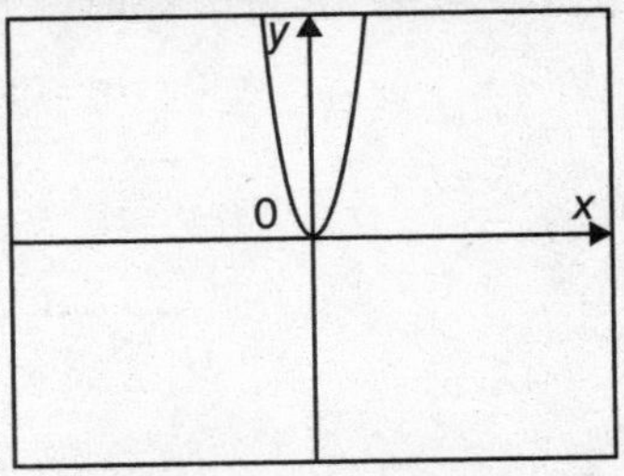
$f(ax)$	Substituting ax for x in $f(x)$, where a is a positive fraction, $0 < a < 1$, results in a horizontal elongation. (For this curve, the effect is the same as a vertical shrink.)	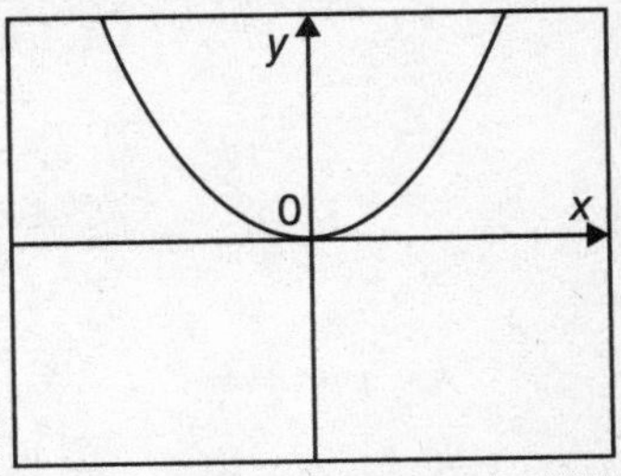
$-f(x)$	The opposite of $f(x)$ is just a reflection across the x-axis.	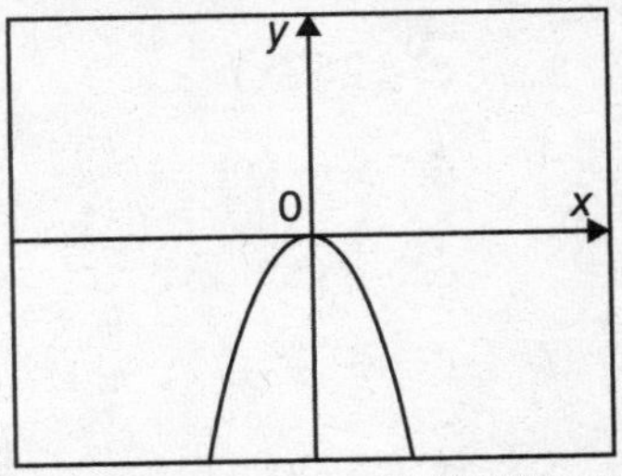
$f(-x)$	Substituting the opposite of x in $f(x)$ results in a reflection across the y-axis. (This curve, because it has y-axis symmetry, looks unchanged.)	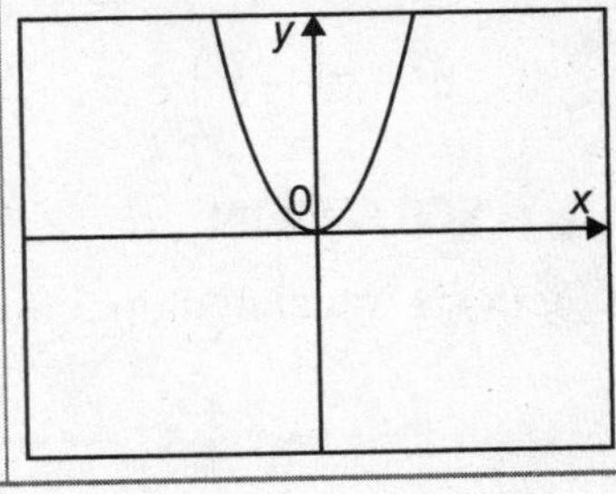

EXAMPLE

Let $g(x)$ be the curve shown in the figure on the next page. Sketch the results of the following transformations.

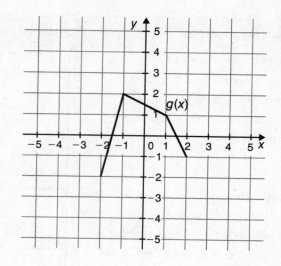

a. $g(x) + 2$

b. $g(x) - 1$

c. $2g(x)$

d. $-g(x)$

e. $g(x - 3)$

f. $g(x + 2) + 1$

g. $g(2x)$

h. $g\left(\dfrac{1}{2}x\right)$

SOLUTION

a. $g(x)$ is translated up 2 units.

b. $g(x)$ is translated down 1 unit.

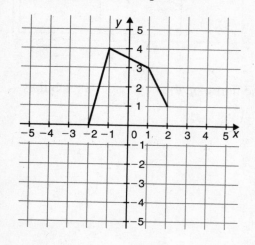

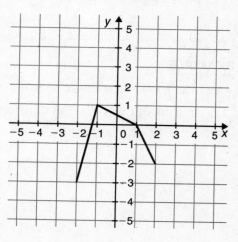

c. $g(x)$ is vertically stretched by a factor of 2. It goes twice as high and twice as low, but the horizontal extent stays the same.

d. $g(x)$ is reflected across the x-axis

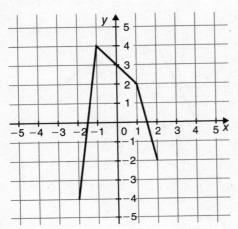

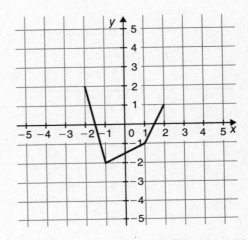

e. $g(x)$ is translated 3 units to the right.

f. This is a combination: $g(x)$ is translated 2 units to the left and 1 unit up.

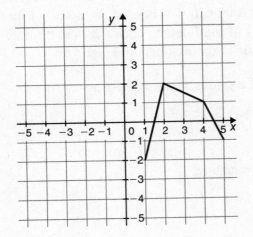

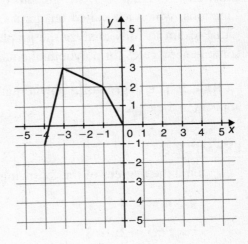

g. $g(x)$ is horizontally compressed by a factor of 2, but the vertical extent stays the same.

h. $g(x)$ is horizontally elongated by a factor of 2, but the vertical extent stays the same.

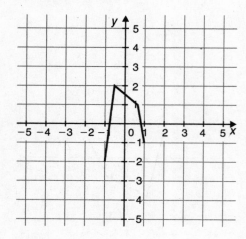
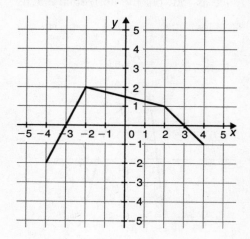

POLYNOMIAL GRAPHS

Earlier in this chapter, we examined the graphs of first-degree, or linear, functions. In Chapter 5, we go into depth about the graphs of second-degree functions, which are called quadratics. Although it is beyond the scope of the CLEP exam to fully analyze the graphs of third-degree equations and beyond, some general rules for polynomial equations are important to know.

Rule 1. The **degree** of a polynomial equation is the highest power of x that appears in the equation.

EXAMPLE

Find the degrees of the following equations.

a. $f(x) = 5$

b. $f(x) = 2x + 3$

c. $f(x) = -x^2 + 8x - 2$

d. $y = 2x^3 + 3x - 2$

e. $f(x) = x - \dfrac{2}{3}x^5$

SOLUTION

a. There is no x, so the degree is zero.

b. The highest power of x is 1, so the degree is 1.

c. The highest power of x is 2, so the degree is 2.

d. The highest power of x is 3, so the degree is 3. (Note that there is no x^2 term; we care only about the *highest* degree.)

e. The highest power of x is 5, so the degree is 5. (Note that the equation is not in descending order of powers; again, we care only about the *highest* degree.)

Rule 2. All polynomial equations graph functions whose domain is $(-\infty, \infty)$. Their graphs pass the vertical line test. The shapes of the graph will be curves (except for linear functions) that are smooth and continuous (i.e., they have no breaks) and have no sharp corners.

EXAMPLE

Which of the following graphs are polynomials? If it is not a polynomial, explain why not.

a.
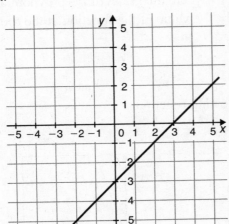

b.
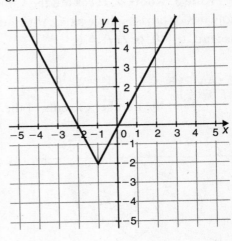

c.

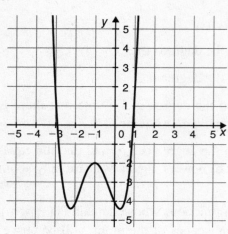

d.

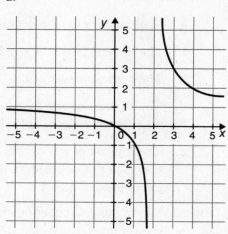

SOLUTION

a. A line is a polynomial.

b. This is not a polynomial since it has a sharp corner.

c. A smooth curve is a polynomial.

d. This is not a polynomial since it is not continuous; it has a break in it.

Rule 3. Roots or **zeros** are the values of x where the graph of the polynomial touches or crosses the x-axis. Algebraically, the value of the function at the root is zero, $y = 0$, or $f(x) = 0$.

EXAMPLE

Find the number of roots of the following functions.

a.

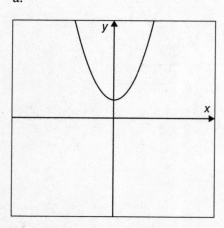

b.

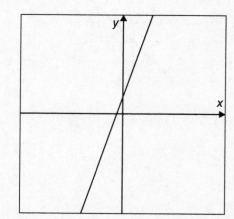

c.
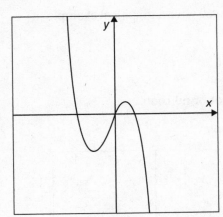

d.
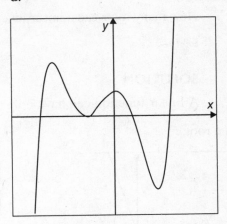

SOLUTION

a. The graph never crosses the *x*-axis, so it has no roots.

b. The graph crosses the *x*-axis once, so it has one root.

c. The graph crosses the *x*-axis three times, so it has three roots.

d. The graph crosses the *x*-axis three times and "bounces" off the *x*-axis once, so it has four roots.

EXAMPLE

If $x = -4$ is a root of $x^3 + 2x^2 + kx + 4$, what is the value of k?

SOLUTION

Substitute $x = -4$. If -4 is a root, the value of the function at $x = -4$ must be zero:

$(-4)^3 + 2(-4)^2 + k(-4) + 4 = 0$

$-64 + 32 - 4k + 4 = 0$

$4k = -28 \Rightarrow k = -7$.

Rule 4. The **maximum number of real roots** that a polynomial function may have is the degree of the polynomial. (The formal definition of "real" is given in Chapter 6, but for now we consider it as the number of places the function crosses the *x*-axis.)

EXAMPLE

How many roots can a linear function have? Give a graphical example of each.

SOLUTION

A linear function can have only one root or no roots.

One root: No roots:

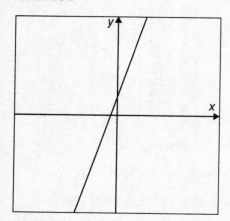 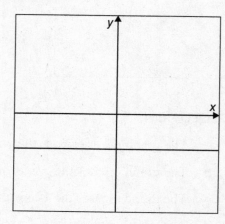

EXAMPLE

A second-degree equation (called a **quadratic**) has the shape in the figure below. Show all the possibilities for the number of roots such an equation can have.

SOLUTION

Two roots:

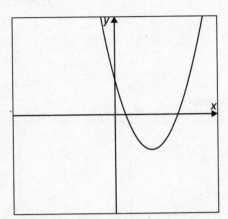

One root:

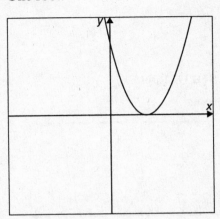

No roots:

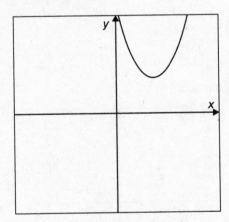

Rule 5. The **minimum number of real roots** that a polynomial function may have depends on the degree of the polynomial. If the degree is an even number, it can have ≥ 0 real roots. If the degree is an odd number, it must have ≥ 1 real root (at least one real root).

EXAMPLE

A third-degree equation (called a **cubic**) has the shape in the figure on the next page. Show all the possibilities for the number of roots such an equation can have.

SOLUTION

Three roots:

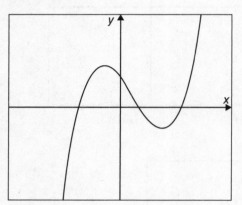

Two roots:

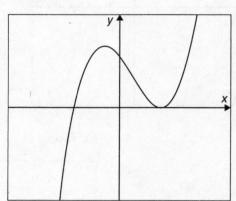

One root:

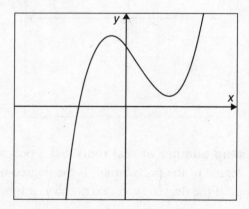

Rule 6. When a polynomial is in factored form,

$$P(x) = (x - a)(x - b)(x - c) \ldots,$$

each one of the factors set equal to zero will create a real root. If the factor is used only once, the root is said to have a **multiplicity** of one. If a factor is used more than once (i.e., it is raised to a power), then the multiplicity of that root will be the power of that factor.

CHAPTER 4: FUNCTIONS AND THEIR PROPERTIES

EXAMPLE
Find the number and location of the roots of $f(x) = x(x - 2)(x + 5)$.

SOLUTION
The roots are $x = 0$, $x - 2 = 0 \Rightarrow x = 2$, and $x + 5 = 0 \Rightarrow x = -5$.
This function has three roots, each with a multiplicity of one.

EXAMPLE
Find the number and location of the roots of $f(x) = (x + 3)(x - 1)^2(x - 4)$.

SOLUTION
The roots are $x + 3 = 0 \Rightarrow x = -3$; $(x - 1)^2 = 0 \Rightarrow x = 1$;
$x - 4 = 0 \Rightarrow x = 4$. This function has three roots. Roots $x = -3$ and $x = 4$ each have a multiplicity of one, and $x = 1$ has a multiplicity of two.

Rule 7. When a polynomial $P(x)$ evaluated at $x=a$ gives a positive number and evaluated at $x = b$ gives a negative number, there must be at least one root between a and b.

EXAMPLE
Determine whether the polynomial $f(x) = x^3 + 3x^2 - 19x + 6$ has a real root between the following integers:

a. 0 and 1
b. 1 and 2
c. 2 and 3

SOLUTION

a. $f(0) = 6$, $f(1) = 1 + 3 - 19 + 6 = -9$. Since the function switches signs, there is a real zero between 0 and 1.

b. $f(1) = -9$, $f(2) = 8 + 12 - 38 + 6 = -12$. Since the function does not switch signs, there does not have to be a real zero between 1 and 2.

c. $f(2) = -12$, $f(3) = 27 + 27 - 57 + 6 = 3$. Since the function switches signs, there is a real zero between 2 and 3.

Rule 8. When an expression with degree two does not factor, it creates two **imaginary roots**. (The formal definition of imaginary numbers is given in Chapter 6, but here we simply say that imaginary roots are not locations where the function crosses the x-axis.)

EXAMPLE

Explain the nature of the roots of $f(x) = x^2 + 4$.

SOLUTION

This is a function of degree two that has a maximum of two roots (see rule 4). Since this function, the sum of squares, is prime, it does not factor. Therefore, both roots must be imaginary.

EXAMPLE

Explain the nature of the roots of $f(x) = (x - 1)(x^2 + 4)$.

SOLUTION

This is a function of degree three that has a maximum of three roots. Since $x^2 + 4$ is prime, two roots must be imaginary. However, for the first factor, $x - 1 = 0 \Rightarrow x = 1$. So this function has one real root and two imaginary roots.

Rule 9. For a polynomial function $P(x)$, if the remainder when $P(x)$ is divided by $x - a$ equals zero (that is, $x - a$ divides evenly into $P(x)$), then a is a root of $P(x)$.

EXAMPLE

Show that $x = 2$ is a root of $f(x) = x^2 + x - 6$.

SOLUTION

$2^2 + 2 - 6 = 4 + 2 - 6 = 0$, so 2 is a root.

Alternatively, $\dfrac{x^2 + x - 6}{x - 2} = \dfrac{(x+3)(x-2)}{x-2} = x + 3$, and since $x - 2$ divides evenly into the polynomial with no remainder, 2 is a root.

Rule 10. For a polynomial function $P(x)$, the function value $P(a)$ is the remainder when $P(x)$ is divided by $x - a$. This is known as the **remainder theorem**.

EXAMPLE

Find the remainder when $x^3 - x^2 + 2x - 5$ is divided by $x - 2$.

SOLUTION

Since we generally want to avoid long division, by using the remainder theorem (rule 10) this remainder is found by evaluating the expression $x^3 - x^2 + 2x - 5$ at $x = 2$: $2^3 - 2^2 + 2(2) - 5 = 8 - 4 + 4 - 5 = 3$. So the remainder is 3.

EXAMPLE

Find the remainder when $x^{50} - 2x^{25} - 4x + 3$ is divided by $x + 1$.

SOLUTION

We really don't want to do long division on this 50th-degree polynomial, so we use rule 10 to get the remainder by evaluating $x^{50} - 2x^{25} - 4x + 3$ at $x = -1$: $(-1)^{50} - 2(-1)^{25} - 4(-1) + 3 = 1 - 2(-1) + 4 + 3 = 10$. So the remainder is 10.

INVERSE FUNCTIONS

The topic of inverses in math confuses many students. The **inverse of a function** is written as f^{-1}. Students mistakenly believe that since $x^{-1} = \frac{1}{x}$, then $f^{-1} = \frac{1}{f}$. This is decidedly incorrect.

If (x, y) is a point on a function f, then the point (y, x) is on the inverse function, f^{-1}. It is possible, however, that $f^{-1}(x)$ is not a function, even though $f(x)$ is.

EXAMPLE

If $f = \{(2, 3), (3, 5), (-4, 8), (-1, -1), (0, 4)\}$, find f^{-1} and determine whether f^{-1} is a function.

SOLUTION

For the inverse, simply reverse the points: $f^{-1} = \{(3, 2), (5, 3), (8, -4), (-1, -1), (4, 0)\}$. This is a function because it has no x-value repeats.

If a function is given in *equation* form, to find the inverse, replace all occurrences of x with y and all occurrences of y with x. Then, if possible, solve for y by putting all the y terms on one side of the equation. Note that because we are switching the x's and y's, the domain of f^{-1} is the range of f and the range of f^{-1} is the domain of f.

EXAMPLE

Find the inverse of the function $y = 4x - 3$.

SOLUTION

Inverse: $x = 4y - 3$, so $4y = x + 3$, and $y = \dfrac{x+3}{4}$ is the inverse.

EXAMPLE

Find the inverse of the function $y = \dfrac{4x+5}{x-1}$.

SOLUTION

The inverse of $y = \dfrac{4x+5}{x-1}$ is $x = \dfrac{4y+5}{y-1}$, or $xy - x = 4y + 5$.

Then $xy - 4y = x + 5$

$y(x - 4) = x + 5$

$y = \dfrac{x+5}{x-4}$.

EXAMPLE

Without actually finding the inverse, find the domain and range of the inverse of the function $f(x) = \sqrt{x-2} + 3$.

SOLUTION

The domain and range are simply switched. For $f(x)$, the domain is $[2, \infty)$, and the range is $[3, \infty)$ Therefore, for the inverse $f^{-1}(x)$, the domain is $[3, \infty)$, and the range is $[2, \infty)$.

EXAMPLE

If the graph of f is as shown below, draw f^{-1} on the same figure and determine whether f^{-1} is a function.

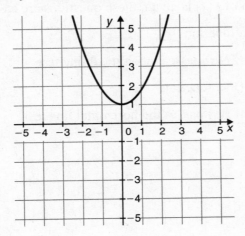

SOLUTION

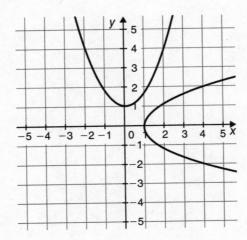

Since the function goes through the points (0, 1), (1, 2), (2, 4), (−1, 2), (−2, 4) the inverse passes through the points (1, 0), (2, 1), (4, 2), (2, −1), (4, −2). Since the x values repeat and since the graph fails the vertical line test, the inverse is not a function. We could also look at the graph of f and realize that since it fails the **horizontal line test**, its inverse is not a function. If we draw a horizontal line through the graph, there are locations where the line intersects the graph in more than one point.

When a graph of $f(x)$ is given, these questions are easily answered:

Is $f(x)$ a function?

Yes, if $f(x)$ passes the vertical line test.

Is $f^{-1}(x)$ a function?

Yes, if $f(x)$ passes the horizontal line test.

CHAPTER 5

Equations and Inequalities

CHAPTER 5

EQUATIONS AND INEQUALITIES

ABSOLUTE VALUE EQUATIONS AND INEQUALITIES

The definition of an **absolute value** of a number is its distance from zero. An absolute value is never a negative number. The formal definition of an absolute value is a piecewise function: $|a| = \begin{cases} a, & a \geq 0 \\ -a, & a < 0 \end{cases}$.

For example, $|8| = 8$ and $|-8| = -(-8) = 8$.

Absolute Value Equations

From the above definition of absolute value, a linear equation involving absolute values can actually be treated as two equations. In the first equation, we replace the absolute value by a positive parenthesis. In the second equation, we replace the absolute value by a negative parenthesis. For example, $|x + 2| = 7$ means either $(x + 2) = 7$ or $-(x + 2) = 7$. We then solve both equations. We must check both solutions in the original equation because this solution process could introduce extraneous values.

EXAMPLE
Solve $|2x - 1| = 15$.

SOLUTION
The two equations and their solutions are

$(2x - 1) = 15 \quad -(2x - 1) = 15$
$2x - 1 = 15 \quad -2x + 1 = 15$
$2x = 16 \quad\quad\quad -2x = 14$
$x = 8 \quad\quad\quad\quad x = -7$

Check with $x = 8$: $|16 - 1| = 15$.
Check with $x = -7$: $|-14 - 1| = 15$.
Both solutions check, and the answer is $x = 8$ or $x = -7$.

EXAMPLE

Solve $|4x - 5| + 5x + 2 = 0$.

SOLUTION

Form two equations as follows:

$4x - 5 + 5x + 2 = 0$ $-(4x - 5) + 5x + 2 = 0$
$9x = 3$ $-4x + 5 + 5x + 2 = 0$
$x = \dfrac{1}{3}$ $x = -7$

Check for $x = \dfrac{1}{3}$: $\left|\dfrac{4}{3} - 5\right| + \dfrac{5}{3} + 2 = \dfrac{11}{3} + \dfrac{5}{3} + 2$, which does not equal zero.

Check for $x = -7$: $|4(-7) - 5| + 5(-7) + 2 = 33 - 35 + 2 = 0$.

Therefore, only $x = -7$ is a solution.

Absolute Value Inequalities

For an inequality involving an absolute value, we replace the absolute value with a positive parenthesis and a negative parenthesis, just as we did for absolute value equations, being sure to preserve the original inequality sign. We then solve each inequality and place each solution on a number line to determine which intervals satisfy the inequality. We then check the solution by substituting into the original inequality a value for x that is in the solution interval.

EXAMPLE:

Solve $|4x - 2| \leq 14$.

SOLUTION:

$(4x - 2) \leq 14$ $-(4x - 2) \leq 14$
$4x - 2 \leq 14$ $-4x + 2 \leq 14$
$4x \leq 16$ $-4x \leq 12$
$x \leq 4$ $x \geq -3$

(Note that since we divided by a negative number (-4) in the right-hand solution, we reversed the inequality.)

```
--------•------------•--------
       -3            4
```

The solution is therefore $-3 \leq x \leq 4$, or $[-3, 4]$ in interval notation.

Check for $x = 0$: $|4(0) - 2| \leq 14 \Rightarrow 2 \leq 14$, which is correct.

EXAMPLE

Solve $|2x - 1| > x + 4$.

SOLUTION

$(2x - 1) > x + 4$ $-(2x - 1) > x + 4$
$2x - 1 > x + 4$ $-2x + 1 > x + 4$
$x > 5$ $-3x > 3$
                    $x < -1$ (Again, we switch the inequality sign here.)

```
←-------o            o------→
       -1            5
```

So $x < -1$ or $x > 5$, or $(-\infty, -1) \cup (5, \infty)$. Note that the two intervals do not overlap in this example, so the answer has two inequalities for x joined by "or." We must check each one separately.

Check for $x = -2$: $|2(-2) - 1| > (-2) + 4 \Rightarrow |-5| > -2 + 4 \Rightarrow 5 > -2 + 4 \Rightarrow 5 > 2$, which is correct.

Check for $x = 6$: $|2(6) - 1| > 6 + 4 \Rightarrow |11| > 10 \Rightarrow 11 > 10$, which is also correct.

QUADRATIC EQUATIONS AND INEQUALITIES

Quadratic Equations

As we saw in Chapter 4, second-degree equations are in the form of $ax^2 + bx + c = 0$ and are called **quadratic equations**. Quadratic functions can have either two roots, one root, or no roots. Since the roots of the quadratic equation indicate where the quadratic function equals zero, our technique for solving quadratic equations is to set the function equal to zero by moving all the terms to one side of the equal sign, leaving 0 on the other side. Then, to simplify the function, factor it, set each factor equal to zero, and solve for x. Alternatively, the solutions for x can be found by using the quadratic formula, introduced in the next section, especially useful when factoring isn't straightforward.

The values of x found in this way are called **solutions** of the equation. Graphically, these are the points where the graph of the function crosses or touches the x-axis (where y or $f(x) = 0$).

EXAMPLE

Show how the graphs of the following quadratic functions match the solutions to the quadratic equations.

a. $f(x) = x^2$
b. $f(x) = x^2 - 4$
c. $f(x) = x^2 + 1$

SOLUTION

a. The graph of this common function, $y = x^2$, as shown in Chapter 4, is the following:

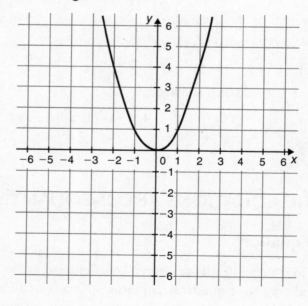

This function is called a **parabola**. It has one root, as seen in the graph. When we set the function equal to zero, we get the equation $x^2 = 0$, which has one solution, $x = 0$, as can be seen on the graph.

b. As discussed in Chapter 4, $y = x^2 - 4$ is the function $y = x^2$ shifted down 4 units. The graph is as follows:

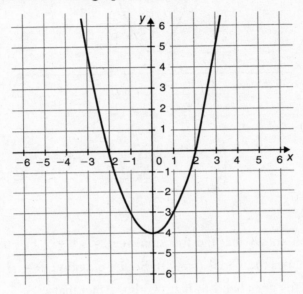

The graph shows that the function $f(x) = x^2 - 4$ has two roots, so the equation $y = x^2 - 4 = 0$ has two solutions, which can be found by factoring.

$x^2 - 4 = 0$

$(x + 2)(x - 2) = 0$

$x + 2 = 0 \qquad x - 2 = 0$

$x = -2 \quad \text{or} \quad x = 2$

Note that when there are two solutions, both are true and we can connect them with "or."

c. As also discussed in Chapter 4, $y = x^2 + 1$ is the function $y = x^2$ shifted up 1 unit. The graph is as follows:

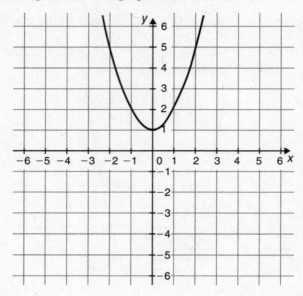

The graph shows that the function $f(x) = x^2 + 1$ has no roots since it never touches the x-axis ($y \neq 0$). So the equation $y = x^2 + 1$ has no solutions. This goes hand-in-hand with the fact that $x^2 + 1$ does not factor. If we set the function equal to zero, we get $x^2 = -1$, which has no real solutions. We will discuss this further in Chapter 6, where we will introduce complex numbers.

EXAMPLE

Solve for x: $x^2 + 3x + 2 = 0$.

SOLUTION

$(x + 2)(x + 1) = 0$
$x + 2 = 0 \quad x + 1 = 0$
$x = -2 \quad \text{or} \quad x = -1$

EXAMPLE

Solve for x: $x^2 - 10x + 25 = 0$.

SOLUTION

$(x - 5)(x - 5) = 0$

$x - 5 = 0$

$x = 5$

This represents a double root since both roots are the same.

EXAMPLE

Solve for x: $4x^2 = 1$.

SOLUTION

Rewrite this equation as $4x^2 - 1 = 0$, which is the difference of two squares.

$(2x + 1)(2x - 1) = 0$

$2x + 1 = 0 \qquad 2x - 1 = 0$

$2x = -1 \qquad 2x = 1$

$x = -\dfrac{1}{2}$ or $x = \dfrac{1}{2}$

EXAMPLE

Solve for x: $2x^2 + 9x = 18$.

SOLUTION

Rewrite the equation as $2x^2 + 9x - 18 = 0$.

$(2x - 3)(x + 6) = 0$

$2x - 3 = 0 \qquad x + 6 = 0$

$2x = 3$

$x = \dfrac{3}{2}$ or $x = -6$

When the equation is more complex, bring all terms to one side, simplify, set the quadratic equation equal to zero, and then attempt to factor.

EXAMPLE

Solve for x: $2x^2 - 7x + 2 = x^2 - 3x + 14$.

SOLUTION

$2x^2 - 7x + 2 - x^2 + 3x - 14 = 0$

$x^2 - 4x - 12 = 0$

$(x + 2)(x - 6) = 0$

$x = -2$ or $x = 6$

EXAMPLE

Solve for x: $(3x - 5)(x + 1) = 2x^2 - 2$.

SOLUTION

$(3x - 5)(x + 1) - 2x^2 + 2 = 0$

$3x^2 - 2x - 5 - 2x^2 + 2 = 0$

$x^2 - 2x - 3 = 0$

$(x - 3)(x + 1) = 0$

$x = 3$ or $x = -1$

If an equation contains a square root, we need to move the radical to one side of the equation and square both sides (see Chapter 3). It is possible that when we perform these operations, the resulting equation may be a quadratic, in which case we proceed as above for any quadratic, but the squaring procedure can introduce extraneous answers, so we must check each answer in the *original* equation.

EXAMPLE

Solve for x: $\sqrt{2x+7} - x = 2$.

SOLUTION

$\sqrt{2x+7} = x + 2$

$\left(\sqrt{2x+7}\right)^2 = (x+2)^2$

$2x + 7 = x^2 + 4x + 4$

$0 = x^2 + 4x + 4 - 2x - 7$ (Notice that it makes no difference which side of the equation has the 0.)

$0 = x^2 + 2x - 3$

$0 = (x - 1)(x + 3)$

$x - 1 = 0 \quad x + 3 = 0$

$x = 1 \quad \text{or} \quad x = -3$

Check for $x = 1$: $\sqrt{2+7} - 1 = 3 - 1 = 2$

Check for $x = -3$: $\sqrt{-6+7} - (-3) = 1 + 3 \neq 2$

So there is only one solution, $x = 1$. The other value, $x = -3$, is extraneous.

The Quadratic Formula

When the quadratic equation is in the form of $ax^2 + bx + c = 0$, it may be difficult or impossible to factor the equation. In that case, we can use the **quadratic formula** to solve the equation. The quadratic formula gives a formula for the solution to the quadratic equation.

$$x = \frac{-b \pm \sqrt{b^2 - 4ac}}{2a}$$

The $\pm$ sign allows two solutions (which may be identical). The **discriminant** $b^2 - 4ac$ tells how many solutions the quadratic equation has:

$b^2 - 4ac \begin{cases} > 0, \text{ 2 real solutions} \\ = 0, \text{ 1 real solution (actually, 2 identical solutions)} \\ < 0, \text{ 0 real solutions (or 2 imaginary solutions; see Chapter 6)} \end{cases}$

EXAMPLE

Solve for x: $x^2 + 3x - 1 = 0$.

SOLUTION

This quadratic is not readily factorable, but it can be solved by using the quadratic formula.

Here, $a = 1, b = 3, c = -1$.

$$x = \frac{-3 \pm \sqrt{3^2 - 4(1)(-1)}}{2(1)} = \frac{-3 \pm \sqrt{9+4}}{2} = \frac{-3 \pm \sqrt{13}}{2}$$

The two solutions are $x = \dfrac{-3 + \sqrt{13}}{2}$ and $x = \dfrac{-3 - \sqrt{13}}{2}$, but we usually leave the answer in the $\pm$ form.

EXAMPLE

Solve for x: $12x^2 = 5x + 2$

SOLUTION

First, set the function equal to zero:

$12x^2 - 5x - 2 = 0$

If we don't want to try to factor the expression, we can use the quadratic formula.

Here, $a = 12, b = -5, c = -2$.

$$x = \frac{-(-5) \pm \sqrt{(-5)^2 - 4(12)(-2)}}{2(12)} = \frac{5 \pm \sqrt{25+96}}{24} = \frac{5 \pm \sqrt{121}}{24} = \frac{5 \pm 11}{24}$$

$$x = \frac{16}{24} = \frac{2}{3} \quad \text{or} \quad x = \frac{-6}{24} = -\frac{1}{4}$$

EXAMPLE

Find the number of solutions of $6x^2 = 8x - 3$.

SOLUTION

$6x^2 - 8x + 3 = 0$

$a = 6, b = -8, c = 3$

Since we are asked to find only the number of solutions, we use the discriminant:

$b^2 - 4ac = (-8)^2 - 4(6)(3) = 64 - 72 = -8$.

Since the discriminant is negative, this quadratic equation has no real solutions.

EXAMPLE

For the quadratic equation $4x^2 + 6x + c = 0$, for what value of c does the equation have two real solutions?

SOLUTION

$a = 4, b = 6$

For the equation to have two real solutions, the discriminant must be positive.

Therefore, $b^2 - 4ac = (6)^2 - 4(4)c > 0$ for two real solutions.

$36 - 16c > 0$

$-16c > -36$

$c < \dfrac{-36}{-16}$, so $c < \dfrac{9}{4}$

(Notice that the inequality is switched because we divided by a negative, –16.)

Quadratic Inequalities

To solve a quadratic inequality, we use a combination of the techniques we used for solving quadratic equations and inequalities. We move all terms to one side of the inequality, preserving the inequality sign. We solve the quadratic by factoring or by using the quadratic formula, and then we create a number line with the solution(s) to determine the solution intervals. As before, it makes sense to check the solution by choosing value(s) in the solution interval(s) and plugging them into the original inequality.

EXAMPLE

Solve for x: $x^2 - 3x > 18$.

SOLUTION

$x^2 - 3x - 18 > 0 \Rightarrow (x + 3)(x - 6) > 0$

For $(x + 3)(x - 6) = 0$, $x = -3, x = 6$.

Since the product is positive (> 0), we need two positive or two negative factors. Our choices are $x < -3$ or $x > 6$, or in interval notation, $(-\infty, -3) \cup (6, \infty)$. (Anything outside of these intervals would allow a positive factor and a negative factor.) The number line therefore looks like:

$$\xleftarrow{\qquad\qquad}\underset{-3}{\circ}\xrightarrow{\qquad\qquad}\underset{6}{\circ}\xrightarrow{\qquad\qquad}$$

Check for $x = -4$: $(-4)^2 - 3(-4) = 16 + 12 = 28 > 18$, which is correct.

Check for $x = 7$: $7^2 - 3(7) = 49 - 21 = 28 > 18$, which is also correct.

EXAMPLE

Find the domain of $\sqrt{32 - 2x^2}$.

SOLUTION

To find the domain of a square root, the radicand (expression under square root) must be ≥ 0. Therefore, $32 - 2x^2 \geq 0$.

$2(16 - x^2) \geq 0$

$2(4 + x)(4 - x) \geq 0$

$4 + x \geq 0 \qquad 4 - x \geq 0$

$x \geq -4 \qquad\quad -x \geq -4$

$\qquad\qquad\qquad x \leq 4$

So $-4 \leq x \leq 4$, or $[-4, 4]$

$$\underset{-4}{\bullet}\rule{3cm}{0.4pt}\underset{4}{\bullet}$$

Check for $x = 0$: $\sqrt{32 - 2(0)^2} = \sqrt{32} \geq 0$, which is correct.

For numbers outside the solution interval, we perform extra checks to show that they are incorrect.

Extra check for $x = 5$: $\sqrt{32 - 2(5)^2} = \sqrt{32 - 50} = \sqrt{-18}$

Extra check for $x = -5$: $\sqrt{32 - 2(-5)^2} = \sqrt{32 - 50} = \sqrt{-18}$.

The checks for values not in the solution interval yield imaginary numbers.

SYSTEMS OF LINEAR EQUATIONS AND INEQUALITIES

Systems of Linear Equations

A **system of linear equations** is a set of two or more linear equations with solutions that are true at the same time. These equations are therefore also called **simultaneous equations**. The system is typically in the form of $\begin{cases} ax + by = c \\ dx + ey = f \end{cases}$.

Since linear equations graph as lines, there are only three possibilities for the two lines.

Case 1. The two lines intersect at a point, as shown in the figure below. These lines intersect at the point (4, 1), so $x = 4$ and $y = 1$ is a solution for both lines. If two distinct lines intersect, they can intersect at only one point. A system with one solution is called **consistent** and independent.

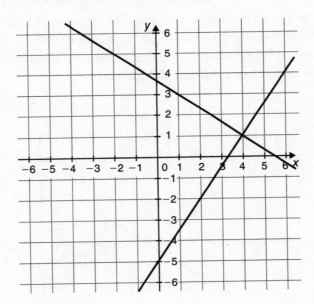

Case 2. The two lines are parallel, as shown in the next figure. Parallel lines never intersect. They will have the same slope but different y-intercepts. A system with no solutions is called **inconsistent**.

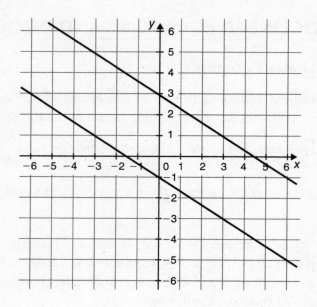

Case 3. The two lines are really only one line, because the equation for one line is a multiple of the other. The solution is an infinite number of points—all the points on the line are solutions as shown in the graph below. A system with infinite solutions is called consistent and **dependent**.

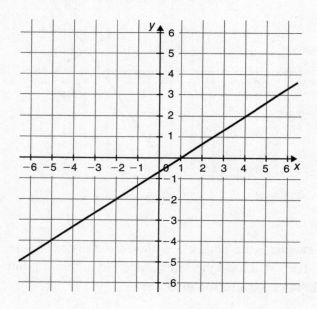

Three methods are available to solve a system of equations: (1) the graphing method, (2) the substitution method, and (3) the elimination method. However, since the CLEP College Algebra exam is a multiple-choice test with five answer choices, rather than going through the somewhat involved methods described next, it is typically easier to substitute each answer choice into the two equations and determine which one satisfies both equations. If the two lines intersect at a point, that is, they are consistent, one of the choices will "work" and the others will not. This "trial-and-error" method works well in a multiple-choice setting and should be high on any list of test-taking strategies. This technique is further explained in Chapter 7.

EXAMPLE

Choose the point that solves the system $\begin{cases} x+y=3 \\ 3x-2y=14 \end{cases}$

(a) (2, 1)   (b) (6, 2)   (c) (5, −2)   (d) (4, −1)   (e) (0, 3)

SOLUTION

We substitute these points until we get one that satisfies both equations.

(a) $\begin{cases} 2+1=3 \\ 6-2=4 \end{cases}$ (b) $\begin{cases} 6+2=8 \\ 18-4=14 \end{cases}$ (c) $\begin{cases} 5-2=3 \\ 15+4=19 \end{cases}$ (d) $\begin{cases} 4-1=3 \\ 12+2=14 \end{cases}$

Thus, solution (d) is correct. Once we get a solution that works, we can stop, so it isn't necessary to check answer choice (e). In fact, it wasn't even necessary to do the second calculation for answer choice (b) because the first part of the calculation for (b) showed it was wrong.

Graphical Method

The **graphical method** involves graphing both equations and finding the intersection. It is quite easy to make errors with this method, and it may involve precise plotting of the lines. The graphical method is not recommended if only points on the lines are given, but if a graph of the lines is given, it should be fairly easy to find the point of intersection and, thus, the solution.

EXAMPLE

Solve the system $\begin{cases} x+y=3 \\ 3x-2y=14 \end{cases}$ graphically.

SOLUTION

First, we put the equations into $y = f(x)$ form. Then we need to graph $y = 3 - x$ and $y = \dfrac{3x}{2} - 7$ by plotting several points for each. The graph looks like the following:

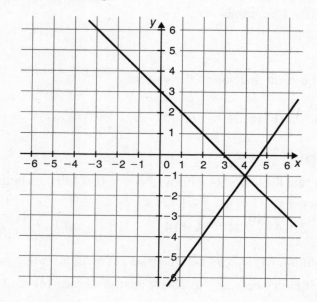

By inspection, the point $(4, -1)$ is the point of intersection, so the solution is $x = 4$, $y = -1$. To be sure this is correct, we can plug the values into both original equations.

How much easier it is to be given the above graph, rather than the equations, and be asked to determine the simultaneous solution!

Substitution Method

The **substitution method** involves using one equation to solve for either variable, and then substituting this expression into the second equation to get the value of other variable. This method is usually best when it is easy to solve for a variable in one of the equations.

EXAMPLE

Solve the system $\begin{cases} x+y=3 \\ 3x-2y=14 \end{cases}$ by the substitution method.

SOLUTION

In the first equation, we solve for x: $x = 3 - y$.

We substitute $x = 3 - y$ into the second equation to get

$3x - 2y = 14$

$3(3-y) - 2y = 14$

$9 - 3y - 2y = 14$

$-5y = 5$

$y = -1$

Now we use the first equation to solve for x.

$x + y = 3$

$x - 1 = 3$

$x = 4$

So the solution is $x = 4$, $y = -1$, which is a solution to both original equations.

Elimination Method

The **elimination method** involves multiplying the equations by numbers that will force the coefficients of one unknown in the resulting equations to be numerically equal but with opposite signs. Then when we add the two resulting equations, that unknown variable cancels out, and we get one equation with one variable, which is easily solvable. We then substitute the value of this variable into either of the original equations to find the second variable.

EXAMPLE

Solve the system $\begin{cases} x+y=3 \\ 3x-2y=14 \end{cases}$ by the elimination method.

SOLUTION

We multiply the first equation by 2 to get a 2y term (which will cancel the $-2y$ term in the second equation), giving the new (equivalent) system:

$\begin{cases} 2x+2y=6 \\ 3x-2y=14 \end{cases}$

Now we add the like terms of both equations:

$5x = 20$

$x = 4$

We plug $x = 4$ into the first equation:

$x + y = 3$

$4 + y = 3$

$y = -1$

So the solution is $x = 4$, $y = -1$.

Again, this is a solution to both original equations.

Which method is best? It depends. If it is easy to solve for a variable, then the substitution method is easier. But in a multiple-choice setting, the trial-and-error process also works well. Unless the graphs are given, the graphical solution takes the most time, so that method is the least desirable.

EXAMPLE

Solve the system $\begin{cases} 8x+3y=5 \\ 5x+4y=18 \end{cases}$

SOLUTION

Since it isn't easy to solve for a variable here, we'll use the elimination method.

If we multiply the first equation by 4 and the second equation by −3 (to eliminate the y variable), we get

$$\begin{cases} 32x + 12y = 20 \\ -15x - 12y = -54 \end{cases}$$

We then add the two equations to get

$17x = -34$

$x = -2$

Plugging $x = -2$ into the first original equation (the second works as well), we get

$-16 + 3y = 5$

$3y = 21$

$y = 7$

So the solution is $x = -2, y = 7$

Check for $(-2, 7)$ in the first equation: $8(-2) + 3(7) = -16 + 21 = 5$, which is correct.

Check for $(-2, 7)$ in the second equation: $5(-2) + 4(7) = -10 + 28 = 18$, which is also correct.

EXAMPLE

Solve the system $\begin{cases} 3x + 5y = 8 \\ 6x + 10y = 12 \end{cases}$

SOLUTION

Since it isn't easy to solve for a variable, again we use the elimination method.

We multiply the first equation by −2 (to eliminate the x variable):

$$\begin{cases} -6x - 10y = -16 \\ 6x + 10y = 12 \end{cases}$$

But when we add the two equations, both variables cancel out and we are left with $0 = -4$, which is clearly wrong. This is an indication that there is no solution to this system (it is inconsistent). Graphically, this would graph as two parallel lines.

It is possible that the CLEP College Algebra exam will have a system of equations with three equations and three variables. For example, $\begin{cases} 2x + 3y - 4z = -8 \\ x + y - 2z = -5 \\ 7x - 2y + 5z = 4 \end{cases}$.

A problem such as this can be solved with the elimination method, but trial-and-error will always be quicker on a multiple-choice exam. For this problem, $x = -1$, $y = 2$, $z = 3$ is the only solution. When we substitute these values into each of the three equations, we get

$2(-1) + 3(2) - 4(3) = -2 + 6 - 12 = -8$

$-1 + 2 - 2(3) = -1 + 2 - 6 = -5$

$7(-1) - 2(2) + 5(3) = -7 - 4 + 15 = 4$

When using trial-and-error, we stop once we get the answer, or we go on to the next answer choice as soon as the given values for x, y, or z yield a wrong total for any of the given equations.

Systems of Linear Inequalities

To solve a system of linear inequalities, it is best to graph each inequality with the proper shading and find the portion of the graph that satisfies both inequalities (the portion that has double shading). Since this takes time, however, the CLEP College Algebra exam usually will give the graph and ask a question about it.

CHAPTER 5: EQUATIONS AND INEQUALITIES | 143

EXAMPLE

Below is a graph of the system of inequalities $\begin{cases} y \leq 4 - \frac{1}{2}x \\ y > 2x - 1 \end{cases}$. If $x = -1$ and y is an integer, how many points satisfy the system?

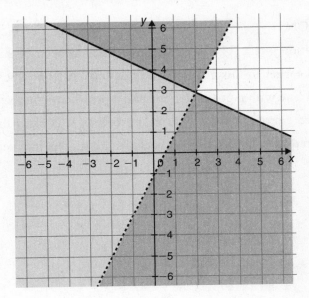

SOLUTION

The solution to the system is the lighter-shaded region. If $x = -1$ and y is an integer, we can read the points $(-1, 4)$, $(-1, 3)$, $(-1, 2)$, $(-1, 1)$, $(-1, 0)$, $(-1, -1)$, and $(-1, -2)$ that satisfy the system from within this region. Note that the point $(-1, -3)$ does not satisfy the system because the dotted line represents greater than ($>$) and not greater than or equal to ($\geq$). Thus, there are seven points with $x = -1$ and y, an integer, that satisfy this system of inequalities.

Advanced Systems of Equations

Systems of equations are not limited to linear functions. A knowledge of basic functions (see Chapter 4) helps with these advanced systems of equations.

EXAMPLE

The system of equations, $\begin{cases} y = x^2 \\ y = ax + b \text{ where } a, b > 0 \end{cases}$

must have how many solutions?

SOLUTION

The equation $y = x^2$ is a basic function that has been discussed in Chapter 4 and earlier in this chapter. The graph of $ax + b$, where $a, b > 0$, is a straight line with a positive slope and a positive y-intercept. The graphs of the two curves look like the figure below. It is clear that there must be two, and only two, solutions because the graphs intersect in only two points.

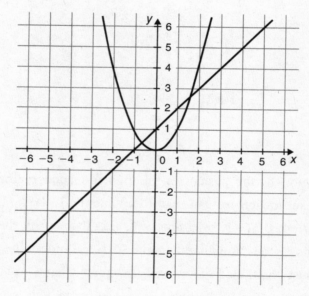

This solution worked because both equations have well-known simple graphs. Problems with unknown graphs for one or both of the equations in the system should be done by trial-and-error for the CLEP College Algebra exam.

CHAPTER 5: EQUATIONS AND INEQUALITIES | 145

EXAMPLE

Which of the following points is a solution to the system $\begin{cases} x^2 + y^2 = 25 \\ x - y = 1 \end{cases}$?

I. $(4, 3)$

II. $(3, 4)$

III. $(-4, -3)$

(a) I only

(b) II only

(c) III only

(d) I and III only

(e) I, II, and III

SOLUTION

By trial-and-error, all of the given points satisfy the first equation, but only the point $(4, 3)$ satisfies $x - y = 1$. The answer is (a).

EXPONENTIAL FUNCTIONS AND LOGARITHMIC EQUATIONS

Exponential Functions

Most of the CLEP College Algebra exam is about linear and polynomial functions, as discussed in Chapter 4 and earlier in this chapter. **Exponential functions**, such as $y = b^x$, also are covered in the exam.

The constant b is called the **base**, where b is a positive number. An exponential graph tends to increase (or decrease) rapidly because the x is in the exponent. The following graphs are examples of exponential curves.

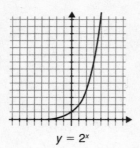
$y = 2^x$

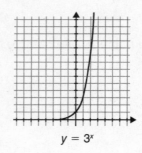
$y = 3^x$

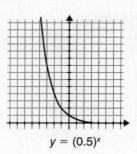
$y = (0.5)^x$

A **growth curve** occurs when $b > 1$, and the larger b is, the steeper the growth curve is. A **decay curve** occurs if $0 < b < 1$, as shown in the third graph above.

All exponential curves in the form of $y = b^x$, whether growth or decay curves, have certain features. The domain of the function is $(-\infty, \infty)$ since the exponent can be any number. The range of exponential functions, however, is $(0, \infty)$ since any positive number raised to a positive power is positive, and a positive raised to a negative power creates a positive fraction. All exponential functions in the form of $y = b^x$ pass through the point $(0, 1)$ because any base b raised to the 0 power equals 1.

We solve basic exponential equations by using the fact that if $b^x = b^y$, then $x = y$. In other words, if the bases are the same, the powers must also be equal.

EXAMPLE

Solve for x: $2^{x+1} = 8$.

SOLUTION

We need to write 8 as a power of 2 so the bases are the same. We know $8 = 2^3$, so we have $2^{x+1} = 2^3$

Thus, $x + 1 = 3$, and $x = 2$.

EXAMPLE

Solve for x: $3^{2x-3} = 9$

SOLUTION

$3^{2x-3} = 3^2$

$2x - 3 = 2$

$2x = 5$

$x = \dfrac{5}{2}$

EXAMPLE

Solve for x: $3^{4x-1} = \dfrac{1}{3}$.

SOLUTION

$3^{4x-1} = 3^{-1}$

$4x - 1 = -1$

$x = 0$

By using the facts for operations with exponents from Chapter 3, we can solve more complicated exponential equations.

EXAMPLE

Solve for x: $4^{3x-3} = 8^{x+2}$.

SOLUTION

We cannot write 8 as a power of 4 but we can write both 4 and 8 as powers of 2.

$(2^2)^{3x-3} = (2^3)^{x+2}$

(Remember that powers raised to powers indicate multiplication of the powers.)

$2^{6x-6} = 2^{3x+6}$

$6x - 6 = 3x + 6$

$3x = 12$

$x = 4$

EXAMPLE

Let f be an exponential function defined by $f(x) = ab^x$, where a and b are both positive constants. If $f(3) = 54$ and $f(5) = 486$, what are the values of a and b?

SOLUTION

$f(3) = 54$ means $ab^3 = 54$.

$f(5) = 486$ means $ab^5 = 486$.

By dividing these two equations, we get

$$\frac{ab^5}{ab^3} = \frac{486}{54} = 9$$

$b^2 = 9$

$b^2 - 9 = 0$

$(b + 3)(b - 3) = 0$

$b = -3$, $b = 3$, but since b must be positive, the answer is only $b = 3$.

To find a, we substitute $b = 3$ into either original function. For example,

$ab^3 = 54 \Rightarrow a(27) = 54$

$a = 2$.

If we used $ab^5 = 486$, we would have the same result:

$ab^5 = 486 \Rightarrow a(243) = 486$

$a = 2$.

The answer is $a = 2$, $b = 3$.

Logarithms

Solving exponential equations such as the ones above are fairly easy when both sides of the equation have common bases. But that technique would not work with an equation such as $3^{x-1} = 6$ since we cannot write 6 as a power of 3. To solve problems of this type, we introduce **logarithms** (also referred to as **logs**). A logarithm is the inverse of an exponential function. We know from Chapter 4, that when we find an inverse, we interchange x and y. So if $y = b^x$, its inverse is $x = b^y$. The use of logarithms helps us to solve exponential functions for y since the notation $\log_b y = x$ is the same as $b^x = y$. The base, b, cannot be negative.

To find the logarithm $\log_b y$, we denote the function as x, or $x = \log_b y$, and then we write the expression as an exponential and we solve it by using the equivalent function above.

EXAMPLE
Find $\log_2 8$.

SOLUTION
We do not know $\log_2 8$, so we write $\log_2 8 = x$.

Now we can write the statement exponentially as $2^x = 8$.

Then $2^x = 2^3$, so $x = 3$, or $\log_2 8 = 3$.

EXAMPLE
Find $\log_3 81$.

SOLUTION
$\log_3 81 = x$

$3^x = 81$

$3^x = 3^4$

$x = 4$, or $\log_3 81 = 4$.

EXAMPLE
Find $\log_4 32$.

SOLUTION
$\log_4 32 = x$

$4^x = 32$

$2^{2x} = 2^5$

$2x = 5$

$x = \dfrac{5}{2}$, or $\log_4 32 = 2.5$.

EXAMPLE

Find $\log_8 \sqrt{2}$.

SOLUTION

$\log_8 \sqrt{2} = x$

$8^x = \sqrt{2}$

$2^{3x} = 2^{\frac{1}{2}}$

$3x = \frac{1}{2}$

$x = \frac{1}{2}\left(\frac{1}{3}\right) = \frac{1}{6}$, or $\log_8 \sqrt{2} = \frac{1}{6}$.

EXAMPLE

Find $\log_9 1$.

SOLUTION

$\log_9 1 = x$

$9^x = 1$

$x = 0$, or $\log_9 1 = 0$.

In fact, the log of 1 for any base is 0 since any term raised to the 0 power equals 1.

Common Logarithms

If the base is not specified, it is assumed to be 10. A logarithm with a base of 10 is called a **common logarithm**. The notations $\log_{10} x$ and $\log x$ are the same.

EXAMPLE
Find log 1000.

SOLUTION
$\log 1000 = x$

Since this is the same as $\log_{10} 1000 = x$,

$10^x = 1000$

$10^x = 10^3$

$x = 3$, or $\log 1000 = 3$.

EXAMPLE
Find log 0.01.

SOLUTION
$\log_{10} 0.01 = x$

$10^x = 0.01 = \dfrac{1}{100} = \dfrac{1}{10^2} = 10^{-2}$

$x = -2$, or $\log 0.01 = -2$

Natural Logarithms

Another number is used as a base in logarithms in many situations. These logarithms are called **natural logarithms**, or **natural logs**, and the base is designated as e. Like π, e is a nonrepeating, nonterminating decimal. Its value is $e = 2.718281828\ldots$ (Note: the 1828 in e, although appearing twice consecutively near the start, does not appear again for a very long while. It is completely coincidental that it appears twice early on.)

Just as common logs use base 10, natural logs (ln) use base e. We just showed that to find the value of a log, we write the expression exponentially. We do the same thing with a natural log, except that now the base is e.

For instance, to find ln 10, call it x, and solve the equation $e^x = 10$. Since e is slightly below 3, we expect ln 10 to be between the values of 2 and 3.

EXAMPLE

Find $\ln e^4$.

SOLUTION

$\ln e^4 = x$

A natural log has a base e, so

$e^x = e^4$

$x = 4$, or $\ln e^4 = 4$.

EXAMPLE

Find $\ln 1$.

SOLUTION

$\ln 1 = x$

$e^x = 1$

$x = 0$

Logarithm Rules

There are three basic rules for operations with logarithms that are important. These rules work with logs to any base or the ln function.

Rule 1. $\log a + \log b = \log(a \cdot b)$.

Rule 2. $\log a - \log b = \log\left(\dfrac{a}{b}\right)$.

Rule 3. $\log a^b = b \log a$.

EXAMPLE

Find the value of $\log 25 + \log 4$.

SOLUTION

Even though we cannot directly find the log of 25 or the log of 4, we can use Rule 1 to say

$\log 25 + \log 4 = \log(25 \cdot 4) = \log 100$

$\log 100 = x$

$10^x = 100$

$10^x = 10^2$

$x = 2$, or $\log 25 + \log 4 = 2$.

EXAMPLE

Find the value of $\log_2 80 - \log_2 5$.

SOLUTION

Even though we cannot directly find $\log_2 80$ or $\log_2 5$, we can use Rule 2 to say

$$\log_2 80 - \log_2 5 = \log_2 \left(\frac{80}{5}\right) = \log_2 16$$

$\log_2 16 = x$

$2^x = 16$

$2^x = 2^4$

$x = 4$, or $\log_2 80 - \log_2 5 = 4$.

EXAMPLE

Find $\log 10^{35}$.

SOLUTION

We can use Rule 3 to say $\log 10^{35} = 35 \log 10$.

$\log 10 = x$

$10^x = 10$

$x = 1$

Since $35(1) = 35$, $\log 10^{35} = 35$.

Logarithmic Equations

Solving **logarithmic equations** (statements that contain logs) is simply a matter of writing them exponentially and solving via the conventional methods discussed earlier in this chapter.

EXAMPLE

Solve for x: $\log_5(2x + 5) = 2$.

SOLUTION

$5^2 = 2x + 5$

$25 = 2x + 5$

$20 = 2x$

$x = 10$

EXAMPLE

Solve for x: $\log_3(4x - 7) = 4$.

SOLUTION

$3^4 = 4x - 7$

$81 = 4x - 7$

$88 = 4x$

$x = 22$

EXAMPLE

Solve for x: $\log_4(x^2 - x + 2) = \dfrac{1}{2}$.

SOLUTION

$4^{1/2} = x^2 - x + 2$.

$2 = x^2 - x + 2$

$0 = x^2 - x$

$x(x - 1) = 0$

$x = 0, x = 1$

(Note that there are two answers because $x^2 - x + 2$ is a quadratic.)

EXAMPLE

Solve for x: $\log_x(2x + 8) = 2$.

SOLUTION

$\log_x(2x + 8) = 2$

$x^2 = 2x + 8$

$x^2 - 2x - 8 = 0$

$(x - 4)(x + 2) = 0$

$x = 4, x = -2$

Since x is the base in the original equation, it cannot be negative, so the only solution is $x = 4$.

EXAMPLE

Solve for x: $\log(x - 1) + \log 4 = 2$.

SOLUTION

We use Rule 1 to rewrite this as $\log 4(x - 1) = 2$ and remember that a log written without a base has base 10.

Then $10^2 = 4(x - 1)$

$100 = 4x - 4$

$104 = 4x$

$26 = x$

CHAPTER 6

Number Systems and Operations

CHAPTER 6

NUMBER SYSTEMS AND OPERATIONS

Chapter 2 informally introduced the concept of sets and mentioned some important sets of numbers. This chapter develops these sets in more detail and defines them more clearly. The first part of this chapter stresses important theory about number systems and operations to give some background for the second part of the chapter, which presents more problems and solutions.

REAL NUMBERS

Number Sets

The set of real numbers (the numbers we have discussed so far in this book) comprise subsets of numbers with different properties. The following discussion defines these subsets of the real number system.

Natural Numbers

The most basic set of real numbers is the set of **natural numbers**, {1, 2, 3, 4, 5,...}. This infinite set is the first set of numbers people learn since it is used to count items.

Closure is an important concept in the set of real numbers. A set is said to be a **closed** set if, when we perform an operation (addition, subtraction, multiplication, or division) on any two numbers in the set, the result is always another number in the set.

For instance, the set of natural numbers is closed under addition. If we take any two natural numbers and add them, we are guaranteed to get another natural number (e.g., $4 + 5 = 9$, $12 + 25 = 37$, $976 + 976 = 1,952$).

However, the set of natural numbers is not closed under subtraction. If we take 4 − 4, we do not get a number in the set of natural numbers, which does not include zero. So mathematicians invented the concept of zero, the absence of a measurable quantity.

Whole Numbers

Adding zero to the set of natural numbers gives us the set of **whole numbers**: $\{0, 1, 2, 3, 4, 5,...\}$. Zero has many applications, such as being flat broke (having no money) or being at sea level (being neither above nor below the average ocean surface). Note that every natural number is a whole number, but not every whole number is a natural number.

Whereas the set of whole numbers is closed under addition (if we add any two whole numbers, we get another whole number), it is still not closed under subtraction. For example, 5 − 6 is not a whole number, nor is 20 − 85.

Integers

So the concept of negative numbers was created—numbers that are less than zero. If we combine the negative numbers with the whole numbers, we then get the set of **integers**, $\{..., -4, -3, -2, -1, 0, 1, 2, 3, 4,...\}$. Of course, we know today that negative numbers have many uses, such as money (owing), temperature (below freezing Centigrade), or altitude (below sea level). Note that every whole number is an integer, but not every integer is a whole number.

With the rules of addition, subtraction, and multiplication of signed numbers established, the set of integers is closed under all three operations. When we add any two integers, we get an integer. When we subtract any two integers, we get an integer. And when we multiply any two integers, we get an integer. But when we come to division, the set of integers is not closed. When we divide any two integers, we are not guaranteed to get an integer. Even though we can sometimes get an integer $\left(\text{e.g., } \dfrac{6}{2}, \dfrac{-20}{4}, \dfrac{-65}{-65}, \dfrac{0}{14}\right)$, there are infinitely many cases for which we don't $\left(\text{e.g., } \dfrac{2}{3}, \dfrac{-3}{8}, \dfrac{15}{-30}, \dfrac{-12}{-15}\right)$.

Rational Numbers

A new number system had to be created to achieve closure under division. These numbers are called **rational numbers**, which are numbers that can be written as fractions in lowest terms. The formal definition of a rational number is $\left\{\dfrac{a}{b}, b \neq 0\right\}$. A rational number is either a fraction or a decimal that terminates (has a definite end) or repeats (has a series of numbers that repeat).

Frequently, it is useful to change decimal numbers to fractions. If the decimal terminates, this is easy, as shown by the following procedure:

1. If there is one decimal place, we place the number without the decimal point over 10 $\left(\text{e.g., } 0.9 = \dfrac{9}{10}\right)$.
2. If there are two decimal places, we place the number without the decimal point over 100 $\left(\text{e.g., } 1.26 = \dfrac{126}{100} = \dfrac{63}{50}\right)$.
3. If there are three decimal places, we place the number without the decimal point over 1000 $\left(\text{e.g., } -0.137 = \dfrac{-137}{1000}\right)$. And so on.

If the decimal repeats, however, the procedure is different: A shorthand to show a decimal repeats is to place a bar over the repeating section. So we write $0.777\ldots$ as $0.\overline{7}$, and 4.838383 as $4.\overline{83}$. Here we focus on decimals for which the repeating part occurs directly after the decimal point. To determine the equivalent fraction, we use the following procedure:

1. If only the first decimal repeats, we place that number (without the decimal point) over 9 $\left(\text{e.g., } 0.444\ldots = 0.\overline{4} = \dfrac{4}{9}\right)$.
2. If the first two decimals repeat, we place that number (without the decimal point) over 99 $\left(\text{e.g., } 0.535353\ldots = 0.\overline{53} = \dfrac{53}{99}\right)$.
3. If the first three decimals repeat, we place that number (without the decimal point) over 999 $\left(\text{e.g., } 0.729729\ldots = 0.\overline{729} = \dfrac{729}{999} = \dfrac{81}{111}\right)$. And so on.

If the repeating decimal occurs in a **mixed number**, which is a number with a whole number part and a decimal part, such as $2.\overline{45}$, we add the whole number to the fraction. Thus, $1.7979... = 1 + 0.\overline{79} = 1 + \frac{79}{99} = \frac{99+79}{99} = \frac{178}{99}$. If the repeating decimal is negative, both parts of the mixed number are negative, for example, $-2.333... = -2 - 0.\overline{3} = -2 - \frac{3}{9} = \frac{-18-3}{9} = \frac{-21}{9} = \frac{-7}{3}$.

There are other techniques to use when the repetition does not start immediately within the decimal, such as for $0.23565656... = 0.23\overline{56}$, but since these won't be tested on the CLEP College Algebra exam, they are beyond the scope of this book.

EXAMPLE

Change the following decimals to fractions in lowest terms.

a. 7.24

b. $-0.555...$

c. $-0.8484...$

d. $1.345345...$

e. $-5.4141...$

SOLUTION

a. $7.24 = \frac{724}{100} = \frac{181}{25}$.

b. $-0.555... = -0.\overline{5} = -\frac{5}{9}$.

c. $0.8484... = 0.\overline{84} = \frac{84}{99} = \frac{28}{33}$.

d. $1.345345... = 1 + 0.\overline{345} = 1 + \frac{345}{999} = \frac{999+345}{999} = \frac{1344}{999}$.

e. $-5.4141... = -5 - 0.\overline{41} = -5 - \frac{41}{99} = \frac{-495-41}{99} = \frac{-536}{99}$.

Irrational Numbers

Irrational numbers are all numbers that are not rational numbers. They *cannot* be written as fractions and include all decimals that never end and never repeat.

Some examples of irrational numbers accurate to five decimal places are the following:

$\pi \approx 3.14159$

$e \approx 2.71828$

$\sqrt{2} \approx 1.4142$

$\sqrt{\frac{1}{5}} \approx .44721$

$\sqrt{3} \approx 1.73205$

$\sqrt[3]{16} \approx 2.51984$

Note the use of the approximate sign ($\approx$). Since these numbers never terminate and never repeat, we cannot write exact decimals to represent them.

Definition of Real Numbers

If we place all rational numbers and all irrational numbers into one set, we get the set of **real numbers**. Formally,

{Real numbers} = { Rational numbers} $\cup$ {Irrational numbers}.

Note that rational numbers and irrational numbers are disjoint sets since a number cannot be both rational and irrational at the same time. Every number we have discussed so far belongs to the set of real numbers.

EXAMPLE

Classify each of the following numbers in all the sets that describe it, using natural numbers, whole numbers, integers, rational numbers, irrational numbers, and real numbers:

a. $\frac{3}{4}$

b. -8

c. $\sqrt{25}$

d. 4π

e. $3.8181\ldots,$

f. 0.243

g. $(-1)^{10}$

h. $3^0 - 1$

i. $\sqrt{45}$

j. $-\dfrac{1182}{3}$

k. $\dfrac{5}{0}$

SOLUTION

	Natural	Whole	Integer	Rational	Irrational	Real
a. $\dfrac{3}{4}$				•		•
b. -8			•	•		•
c. $\sqrt{25} = 5$	•	•	•	•		•
d. 4π					•	•
e. $3.8181\ldots$				•		•
f. $0.243 = \dfrac{243}{1000}$				•		•
g. $(-1)^{10} = 1$	•	•	•	•		•
h. $3^0 - 1 = 1 - 1 = 0$		•	•	•		•
i. $\sqrt{45}$					•	•
j. $-\dfrac{1182}{3} = -394$			•	•		•
k. $\dfrac{5}{0}$ does not exist						

The huge set of real numbers encompasses all numbers that we know to this point. This set is closed under addition, subtraction, multiplication, and division. That is, given two real numbers, we can add them, subtract them, multiply them, and divide them (with the exception of dividing by zero) and be sure that we will get another real number.

What about the operation of squaring? We know that squaring is just another way to write multiplication, so the set of real numbers is closed under squaring as well.

But what about the operation of square roots? We know we can take the square root of any positive number. Some of them are natural, such as $\sqrt{36} = 6$; some are rational, such as $\sqrt{\frac{1}{4}} = \frac{1}{2}$; and some are irrational, such as $\sqrt{7}$. But we cannot take the square root of negative numbers. To this point, $\sqrt{-1}$ is meaningless. Therefore, the set of real numbers is not closed under the operation of square roots.

COMPLEX NUMBERS

Imaginary Numbers

Another set of numbers, called **imaginary numbers**, was invented to include square roots of negative numbers.* We define the **imaginary unit** to be the square root of negative 1, $\sqrt{-1}$, and as a shorthand, we use the symbol i to represent the imaginary unit. So $\sqrt{-1} = i$, and $i^2 = -1$. An imaginary number can be defined as any number in the form of bi, where b is a real number. For instance, the number 3 is real, but the number $3i$ is imaginary.

Having the new concept of i allows us to reexamine the equation $x^2 + 1 = 0$. We found in Chapter 5 that this equation had no solutions because if we square a positive or negative number, we get a positive result and adding 1 will keep the result positive. This is true if we are limited to real solutions. But if we allow ourselves to include imaginary solutions as possibilities, we find that this equation has two imaginary solutions, either i or $-i$, since $i^2 + 1 = -1 + 1 = 0$ and $(-i)^2 + 1 = -1 + 1 = 0$.

---

* How can numbers be "invented"? We saw that zero and negative numbers were created to represent real-world needs. Imaginary numbers also have real-world applications, but they are not covered in the CLEP College Algebra exam, so they are beyond the scope of this book.

We can use the fact that $\sqrt{-a} = \sqrt{-1}\sqrt{a} = i\sqrt{a}$ to simplify square roots of negative numbers. Therefore, $\sqrt{-4} = \sqrt{-1}\sqrt{4} = i\sqrt{4} = 2i$.

EXAMPLE

Find the following:

a. $\sqrt{-49}$

b. $\sqrt{-10}$

c. $\sqrt{-50}$

d. $\sqrt{-48}$

SOLUTION

a. $\sqrt{-49} = \sqrt{-1}\sqrt{49} = i\sqrt{49} = 7i$.

b. $\sqrt{-10} = \sqrt{-1}\sqrt{10} = i\sqrt{10}$.

c. $\sqrt{-50} = \sqrt{-1}\sqrt{50} = i\sqrt{25}\sqrt{2} = 5i\sqrt{2}$.

d. $\sqrt{-48} = \sqrt{-1}\sqrt{48} = i\sqrt{16}\sqrt{3} = 4i\sqrt{3}$.

Definition of Complex Numbers

Putting all the real numbers and all the imaginary numbers together, we get a new set of numbers called **complex numbers**. A complex number has a real part and an imaginary part and takes the form $a + bi$, where a and b are real numbers. Numbers such as $4 + 5i$, $-2 - 3i$, $\frac{1}{2} + 4.7i$, and $\sqrt{3} - i\sqrt{21}$ are all examples of complex numbers.

Actually, *all* numbers are complex. For instance, the number 3 can be written as $3 + 0i$ and the number $-\sqrt{7}$ can be written as $-\sqrt{7} + 0i$. The following figure shows the final relationship of all of the above sets of numbers. Once we establish the most specific subset of a number using this figure, the number will also belong to all sets above it. For example, since -5 is an integer, it is also rational, real, and complex; since $\sqrt{90}$ is irrational, it is also real and complex; since $\sqrt{-11}$ is imaginary, it is also complex.

CHAPTER 6: NUMBER SYSTEMS AND OPERATIONS

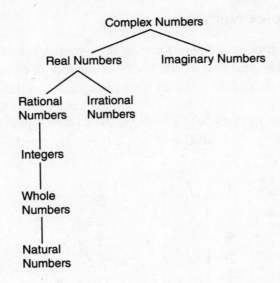

EXAMPLE

Classify the following numbers:

a. $\dfrac{3}{5}$

b. $\sqrt{-100}$

c. 0

d. $\sqrt{\pi}$

e. $10 + \dfrac{7}{2}i$

SOLUTION

a. $\dfrac{3}{5}$ is rational, real, and complex.

b. $\sqrt{-100} = 10i$ is imaginary and complex.

c. 0 is whole, an integer, rational, real, and complex.

d. $\sqrt{\pi}$ is irrational, real, and complex.

e. $10 + \dfrac{7}{2}i$ is just complex.

Plotting Complex Numbers

We can plot points in complex form on the complex set of axes, called the **complex plane**. The real part of the complex number is placed on the x-axis and the imaginary part of the complex number is placed on the y-axis.

The locations of the indicated points on the complex plane graphed below are: $A = 5 + 2i$; $B = -3 + 4i$; $C = -4 - i$; $D = 1$; $E = -5i$.

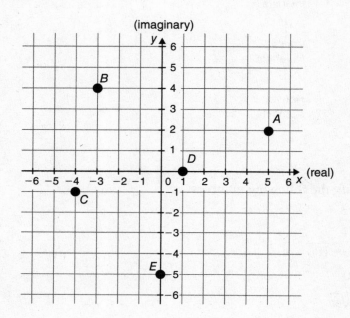

Operations with Complex Numbers

Addition and Subtraction

To add or subtract numbers in complex form, we simply add the real parts and the imaginary parts separately. In symbols, $(a + bi) + (c + di) = (a + c) + (b + d)i$ and $(a + bi) - (c + di) = (a - c) + (b - d)i$.

Thus, $(5 + 3i) + (2 - 4i) = (5 + 2) + (3 - 4)i = 7 - i$.

EXAMPLE

Find the following:

a. $(8 + 2i) + (-6 + i)$

b. $(7 - 8i) - (3i)$

c. $(15 - 7i) - (-15 - 7i)$

d. $(8 - 6i) + (-8 + 5i)$

SOLUTION

a. $(8 + 2i) + (-6 + i) = (8 - 6) + (2 + 1)i = 2 + 3i$.

b. $(7 - 8i) - (3i) = (7 - 0) + (-8 - 3)i = 7 - 11i$.

c. $(15 - 7i) - (-15 - 7i) = (15 - (-15)) + (-7 - (-7))i = 30 + 0i = 30$.

d. $(8 - 6i) + (-8 + 5i) = (8 - 8) + (-6 + 5)i = 0 - i = -i$.

Multiplication

Multiplying two complex numbers utilizes the distributive law and the fact that $i^2 = -1$. When we multiply $4(2 - 3i)$, we multiply the 4 by both terms: $4(2 - 3i) = 8 - 12i$. When we multiply $i(7 + 2i)$, we multiply the i by both terms: $i(7 + 2i) = 7i + 2i^2$. But since $i^2 = -1$, we can say that $i(7 + 2i) = 7i + 2i^2 = 7i - 2 = -2 + 7i$.

EXAMPLE

Find $2(4 - i) - i(1 - i)$.

SOLUTION

$2(4 - i) - i(1 - i)$

$= 8 - 2i - i + i^2$

$= 8 - 3i - 1$

$= 7 - 3i$

When we multiply two complex numbers together, both in the form $a + bi$, we handle it just as we did when we multiplied two binomials, treating it as a FOIL problem. To multiply $(4 + 3i)(2 - 6i)$, we expand it as follows:

F O I L
$8 - 24i + 6i - 18i^2$

$= 8 - 18i - 18(-1)$

$= 8 - 18i + 18$

$= 26 - 18i$

EXAMPLE

Find $(9 - 2i)(4 + i)$.

SOLUTION

$36 + 9i - 8i - 2i^2$

$= 36 + i - 2(-1)$

$= 36 + i + 2$

$= 38 + i$

EXAMPLE

Find $(5 - 3i)^2$.

SOLUTION

$(5 - 3i)^2 = (5 - 3i)(5 - 3i)$

$= 25 - 15i - 15i + 9i^2$

$= 25 - 30i + 9(-1)$

$= 25 - 30i - 9$

$= 16 - 30i$

Division

When we divide two complex numbers, we have to remember that we are not permitted to have a square root in the denominator of a fraction. Since $i = \sqrt{-1}$, we have to eliminate any occurrence of i in the denominator. Thus, for a fraction in the form $\dfrac{a+bi}{ci}$, we multiply both numerator and denominator by i, and that step will eliminate the i from the denominator. For example, to simplify $\dfrac{2+3i}{i}$:

$$\dfrac{2+3i}{i}\left(\dfrac{i}{i}\right)=\dfrac{2i+3i^2}{i^2}$$

$$=\dfrac{2i+3(-1)}{-1}=\dfrac{2i-3}{-1}$$

$$=\dfrac{-3}{-1}+\dfrac{2i}{-1}$$

$$=3-2i$$

EXAMPLE

Simplify $\dfrac{7-2i}{-3i}$.

SOLUTION

$$\dfrac{7-2i}{-3i}\left(\dfrac{i}{i}\right)$$

$$=\dfrac{7i-2i^2}{-3i^2}$$

$$=\dfrac{7i-2(-1)}{-3(-1)}$$

$$=\dfrac{7i+2}{3}$$

For a fraction in the form $\dfrac{a+bi}{c+di}$, we multiply both the numerator and the denominator by the **complex conjugate** of the denominator $(c + di)$, defined as $(c - di)$, and that step will eliminate the i from the denominator. For example, to simplify $\dfrac{3+i}{2+i}$, we multiply both the numerator and the denominator by $2 - i$:

$$\left(\frac{3+i}{2+i}\right)\left(\frac{2-i}{2-i}\right)$$
$$=\frac{6-3i+2i-i^2}{4+2i-2i-i^2}$$
$$=\frac{6-i-(-1)}{4-(-1)}$$
$$=\frac{6-i+1}{4+1}$$
$$=\frac{7-i}{5}$$

EXAMPLE

Simplify $\frac{5+2i}{1-3i}$.

SOLUTION

The complex conjugate in this case is $1+3i$.

$$\left(\frac{5+2i}{1-3i}\right)\left(\frac{1+3i}{1+3i}\right)$$
$$=\frac{5+15i+2i+6i^2}{1+3i-3i-9i^2}$$
$$=\frac{5+17i+6(-1)}{1-9(-1)}$$
$$=\frac{5+17i-6}{1+9}$$
$$=\frac{-1+17i}{10}$$

Solving Quadratic Equations with Imaginary Roots

Let's revisit solving quadratic equations now that we know it is possible also to have imaginary solutions. The easiest way to solve quadratic equations is to use the quadratic formula, which was explained in Chapter 5. To review, the quadratic formula states that for an equation in the form of $ax^2 + bx + c = 0$, $x = \frac{-b \pm \sqrt{b^2 - 4ac}}{2a}$.

Chapter 5 also explained that the **discriminant** $b^2 - 4ac$ tells how many solutions the quadratic equation has:

$$b^2 - 4ac \begin{cases} > 0, & \text{2 real solutions} \\ = 0, & \text{1 real solution (a double solution)} \\ < 0, & \text{0 real solutions (or 2 imaginary solutions)} \end{cases}$$

The important fact here is that quadratic equations always have two solutions, but now we know if there are no real solutions (if $b^2 - 4ac < 0$), the two solutions must be imaginary. For example, to find the number of solutions of the equation $x^2 + 4 = 0$, we check the discriminant, where $a = 1$, $b = 0$, $c = 4$, and $b^2 - 4ac = 0 - 4(4) = -16$. This tells us we have no real solutions and two imaginary solutions.

To find the solutions, we use the quadratic formula:

$$x = \frac{-0 \pm \sqrt{0^2 - 4(1)(4)}}{2(1)} = \frac{\pm\sqrt{-16}}{2} = \frac{\pm\sqrt{-1}\sqrt{16}}{2} = \frac{\pm 4i}{2} = \pm 2i.$$

EXAMPLE

Solve the quadratic equation $x^2 - 3x + 5 = 0$.

SOLUTION

$a = 1$, $b = -3$, $c = 5$

$$x = \frac{-(-3) \pm \sqrt{(-3)^2 - 4(1)(5)}}{2(1)} = \frac{3 \pm \sqrt{9 - 20}}{2} = \frac{3 \pm \sqrt{-11}}{2}$$

$$= \frac{3 \pm \sqrt{-1}\sqrt{11}}{2} = \frac{3 \pm i\sqrt{11}}{2}$$

EXAMPLE

Solve the quadratic equation $3x^2 + 2x + 1 = 0$.

SOLUTION

$a = 3, b = 2, c = 1$

$$x = \frac{-2 \pm \sqrt{2^2 - 4(3)(1)}}{2(3)} = \frac{-2 \pm \sqrt{4-12}}{6} = \frac{-2 \pm \sqrt{-8}}{6}$$

$$= \frac{-2 \pm \sqrt{-1}\sqrt{4}\sqrt{2}}{6} = \frac{-2 \pm 2i\sqrt{2}}{6} = \frac{-1 \pm i\sqrt{2}}{3}$$

SEQUENCES AND SERIES

General Sequences

Consider the set of numbers 4, 7, 10, 13, …. This set of numbers is called a **sequence**. The three dots mean that there are more numbers in the sequence. A sequence that does not end and continues forever is called an **infinite sequence**.

Each number is called a **term** of the sequence. The first term, named a_1, is 4; the second term, named a_2, is 7; the third term, named a_3, is 10; and so on.

Sequences generally have a rule that describes the **nth term** or **general term**. The above sequence could be described as 4, 7, 10, 13, …, $3n + 1$,… or $a_n = 3n + 1$, where n indicates the placement of the term. We find the terms by plugging the numbers $n = 1, 2, 3, …$ into the general term.

EXAMPLE

For the sequences below, find the first three terms, the 10th term, and the 25th term.

a. $a_n = 5n - 5$,

b. $a_n = n^2 - 10n$,

c. $a_n = (-1)^n \frac{n-1}{n+1}$.

CHAPTER 6: NUMBER SYSTEMS AND OPERATIONS

SOLUTION

a. $a_1 = 0, a_2 = 5, a_3 = 10, a_{10} = 45, a_{25} = 120$.

b. $a_1 = -9, a_2 = -16, a_3 = -21, a_{10} = 0, a_{25} = 375$.

c. $a_1 = 0, a_2 = \dfrac{1}{3}, a_3 = \dfrac{-2}{4}, a_{10} = \dfrac{9}{11}, a_{25} = \dfrac{-24}{26}$.

Finding the general term of a sequence involves looking for a pattern, creating a rule, and checking out the rule by plugging in numbers to check it for accuracy. If the terms alternate in signs, we start the sequence with either $(-1)^n$ if the odd-numbered terms are negative, or $(-1)^{n+1}$ if the even-numbered terms are negative. Since the CLEP College Algebra exam is multiple-choice, we often can just plug numbers into each choice to see which answer is correct, rather than trying to figure out the formula.

EXAMPLE

For the sequence, $0, -1, 8, -27, 64$, which of the following is the general term?

(A) $a_n = (-1)^n(n^2 - 1)$

(B) $a_n = n^3$

(C) $a_n = (-1)^{n+1}(n^3 - 1)$

(D) $a_n = (-1)^{n+1}(n - 1)^3$

(E) $a_n = (-1)^n(n - 1)^3$

SOLUTION

For a problem of this type, we can use trial-and-error, remembering that as soon as a choice is ruled out as the answer, we go right to the next choice. We want the sequence $0, -1, 8, -27, 64$. So, for (A), we get $a_1 = (-1)(0) = 0$, but $a_2 = (1)(3) \neq -1$, so we go right to choice (B). We recognize that choice (B) won't give the alternating signs in the sequence, or we can calculate $a_1 = 1 \neq 0$. In either case, (B) isn't correct, so we go right to (C), which has $a_1 = (1)(0) = 0$ but $a_2 = (-1)(7) \neq -1$, so now we go to choice (D), for which all the terms match the original sequence. So (D) is the correct answer.

General Series

A **series** is the sum of a sequence. Because sequences can have an infinite number of terms, we first consider the easier problem of finding a partial sum for a finite sequence. The partial sum of a sequence of the first n terms is the series $S_n = a_1 + a_2 + a_3 + \ldots + a_n$. To find S_n, we simply generate the terms and add them.

EXAMPLE

For $a_n = 2n + 1$, find $S_1, S_2, S_3,$ and S_4.

SOLUTION

The sequence is $a_n = 3, 5, 7, 9, \ldots$.

$S_1 = 3$

$S_2 = 3 + 5 = 8$

$S_3 = 3 + 5 + 7 = 15$

$S_4 = 3 + 5 + 7 + 9 = 24$

EXAMPLE

For $a_n = \dfrac{(-1)^{n+1}}{2^n}$, find $S_1, S_2, S_3,$ and S_4.

SOLUTION

The sequence is $a_n = \dfrac{1}{2}, -\dfrac{1}{4}, \dfrac{1}{8}, -\dfrac{1}{16}, \ldots$

$S_1 = \dfrac{1}{2}$

$S_2 = \dfrac{1}{2} - \dfrac{1}{4} = \dfrac{1}{4}$

$S_3 = \dfrac{1}{2} - \dfrac{1}{4} + \dfrac{1}{8} = \dfrac{3}{8}$

$S_4 = \dfrac{1}{2} - \dfrac{1}{4} + \dfrac{1}{8} - \dfrac{1}{16} = \dfrac{5}{16}$

We use the Greek letter sigma (Σ) to simplify notation for the formula for the general term of a series. The expression "$\sum_{n=1}^{4}(4n-3)$" is read as: "the sum of $4n-3$ as n goes from 1 to 4." So $\sum_{n=1}^{4}(4n-3) = 1+5+9+13 = 28$. Using **sigma notation** won't help to calculate a series any faster, but it is a more convenient way to write it.

EXAMPLE

Compute the following:

a. $\sum_{n=1}^{5}(8-6n)$

b. $\sum_{n=1}^{4} 3^{(n-1)}$

c. $\sum_{n=1}^{4}(-1)^{n+1}\frac{1}{n^2}$

SOLUTION

a. $\sum_{n=1}^{5}(8-6n) = 2-4-10-16-22 = -50$.

b. $\sum_{n=1}^{4} 3^{(n-1)} = 1+3+9+27 = 40$.

c. $\sum_{n=1}^{4}(-1)^{n+1}\frac{1}{n^2} = 1 - \frac{1}{4} + \frac{1}{9} - \frac{1}{16} = \frac{115}{144}$.

The sum of an infinite series is written as $\sum_{n=1}^{\infty} a_n$. For example, the series $1 + \frac{1}{2} + \frac{1}{3} + \frac{1}{4} + \ldots$ is written as $\sum_{n=1}^{\infty}\left(\frac{1}{n}\right)$.

EXAMPLE

Describe the series $\sum_{n=1}^{\infty}\left(\frac{1}{\sqrt{n}}\right)$.

SOLUTION

$$\sum_{n=1}^{\infty}\left(\frac{1}{\sqrt{n}}\right) = 1 + \frac{1}{\sqrt{2}} + \frac{1}{\sqrt{3}} + \frac{1}{\sqrt{4}} + \dots$$

Arithmetic Sequences

A sequence in which a constant d is added to each term to get the next term is called an **arithmetic sequence**. The constant d is called the **common difference**. The common difference d can be positive or negative.

In the sequence 7, 10, 13, 16, 19 ..., $a_1 = 7$ and the common difference d is 3. The common difference is found by subtracting any term from the term that follows it.

EXAMPLE

In the following arithmetic sequences, find the first term a_1 and the common difference d.

a. 5, 11, 17, 23, ...

b. 28, 19, 10, 1, −8, ...

c. 2, $-\frac{1}{2}$, −3, $-\frac{11}{2}$, ...

SOLUTION

a. $a_1 = 5, d = 6$.

b. $a_1 = 28, d = -9$.

c. $a_1 = 2, d = -\frac{5}{2}$.

The **general formula for an arithmetic sequence** is

$$a_n = a_1 + (n-1)d.$$

This formula allows us to find the nth term in an arithmetic sequence if we are given the first term and the common difference d. The difference d is multiplied $(n - 1)$ times because we already have the first term, a_1. Remember this formula for the CLEP College Algebra exam because it is used in several contexts.

For example, to find the 12th term of 7, 15, 23, 31, ... we use
$a_1 = 7, d = 8, n = 12$

so $a_{12} = 7 + (12 - 1)(8)$

$= 7 + 11(8)$

$= 7 + 88$

$= 95.$

EXAMPLE
Find the 25th term in the sequence 11, 23, 35, ...

SOLUTION
$a_1 = 11, d = 12,$ and $n = 25$

so $a_{25} = 11 + (25 - 1)(12)$

$= 11 + 24(12)$

$= 11 + 288$

$= 299.$

EXAMPLE
In the sequence 126, 119, 112, ..., find the 20th term.

SOLUTION
$a_1 = 126, d = -7, n = 20$

so $a_{20} = 126 + (20 - 1)(-7)$

$= 126 + 19(-7)$

$= 126 - 133$

$= -7.$

Conversely, if we know the first term, the common difference d, and the nth term, we can find the number of terms n in the sequence up to (and including) the nth term. For instance, if the sequence is 5, 8, 11, ..., we can find which term has a value of 302. By using the same formula $a_n = a_1 + (n - 1)d$, we get:

$302 = 5 + (n - 1)(3)$

$302 = 5 + 3n - 3$

$3n = 300$

$n = 100$

So 302 is the 100th term.

EXAMPLE

How many terms are in the finite arithmetic sequence 4, 15, 26, ..., 521?

SOLUTION

$a_1 = 4$, $d = 11$, $a_n = 521$, and we must find the value of n.

$521 = 4 + (n - 1)(11)$

$521 = 4 + 11n - 11$

$11n = 528$

$n = 48$

Also, if we know the first term and the last term of a sequence and how many terms there are, we can find the common difference d. We again rely on the arithmetic sequence formula, $a_n = a_1 + (n - 1)d$.

EXAMPLE

If $a_1 = 5$, and $a_{25} = 173$, find the common difference d in the sequence.

SOLUTION

$a_1 = 5$, $a_n = 173$, $n = 25$

$173 = 5 + (25 - 1)d$

$173 = 5 + 24d$

$24d = 168$

$d = 7$

Arithmetic Means

The numbers $m_1, m_2, m_3, \ldots, m_n$ are called the **arithmetic means** between two numbers a and b if $a, m_1, m_2, m_3, \ldots, m_n, b$ forms an arithmetic sequence. Arithmetic means are simply numbers that are inserted between two terms to create an arithmetic sequence. For instance, the single arithmetic mean between 20 and 56 is 38. So 20, 38, 56, ... is a three-term arithmetic sequence with $d = 18$. To find arithmetic means, we again use the arithmetic sequence formula, $a_n = a_1 + (n - 1)d$.

EXAMPLE

Insert three arithmetic means between 7 and 15.

SOLUTION

$a_1 = 7, a_n = 15, n = 5$ (the first and last terms plus the three terms between)

$a_n = a_1 + (n - 1)d$

$15 = 7 + (5 - 1)d$

$15 = 7 + 4d$

$8 = 4d$

$d = 2$

Therefore, the three arithmetic means are 9, 11, and 13, and the sequence is 7, 9, 11, 13, 15.

EXAMPLE

Insert four arithmetic means between 23 and 118.

SOLUTION

$a_1 = 23, a_n = 118, n = 6$ (the first and last terms plus the four terms between)

$a_n = a_1 + (n - 1)d$

$118 = 23 + (6 - 1)d$

$118 = 23 + 5d$

$95 = 5d$

$d = 19$

Therefore, the three arithmetic means are 42, 61, 80, and 99, and the sequence is 23, 42, 61, 80, 99, 118.

Arithmetic Series

We add up the first n terms of an arithmetic sequence to find the **arithmetic series** S_n. Two formulas can help us do the arithmetic without generating all the terms.

If we know the first term a_1, the common difference d, and the value of n, we use:

$$S_n = \frac{n}{2}[2a_1 + (n-1)d].$$

If we know the first term a_1, the nth term a_n, and the value of n, we can use the simpler formula:

$$S_n = \frac{n}{2}(a_1 + a_n).$$

Remember these formulas for the CLEP College Algebra exam because they are used in several contexts. However, if n is small, it is usually easier to write the n terms and just add them.

EXAMPLE

Find the sum of the first 50 terms of $4 + 10 + 16 + \ldots$.

SOLUTION

We know the first term, the common difference, and the number of terms, so we use the formula:

$$S_n = \frac{n}{2}[2a_1 + (n-1)d]$$

$a_1 = 4, d = 6, n = 50$

$$S_{50} = \frac{50}{2}[2(4) + (50-1)(6)]$$

$= 25[8 + 49(6)]$

$= 25(302)$

$= 7550$

EXAMPLE
Find the sum of the numbers $1 + 2 + 3 + \ldots + 100$.

SOLUTION
We know the first term, the last term, and that there are 100 terms, so we can use the formula $S_n = \dfrac{n}{2}(a_1 + a_n)$.

$a_1 = 1, a_n = 100, n = 100$

$S_{100} = \dfrac{100}{2}(1 + 100)$

$= 50(101)$

$= 5050$

For this particular example, we also know the difference d is 1, so we could have used the other formula, $S_n = \dfrac{n}{2}[2a_1 + (n-1)d]$, as well.

$S_{100} = \dfrac{100}{2}[2(1) + (99)(1)]$

$= 50(101)$

$= 5050$

EXAMPLE
Find $\sum\limits_{n=3}^{6}(2n - 1)$.

SOLUTION
This problem involves only four terms, so we just write them out and add them. For $n = 3$, we get $2(3) - 1 = 5$. The sum for $n = 3, 4, 5, 6$ is $5 + 7 + 9 + 11 = 32$.

EXAMPLE

Find $\sum_{n=1}^{33}(100-5n)$.

SOLUTION

The first term is $100 - 5 = 95$. The nth term (for $n = 33$) is $100 - 5(33) = -65$. We can use the formula: $S_n = \frac{n}{2}(a_1 + a_n)$.

$a_1 = 95, a_n = -65, n = 33$

$S_{33} = \frac{33}{2}(95 - 65)$

$= \frac{33}{2}(30)$

$= 495$

EXAMPLE

Ray drives 1000 km in January, 1260 km in February, and an additional 260 km every month for the remainder of the year. How far does he travel during the year?

SOLUTION

$a_1 = 1000, d = 260, n = 12$

$S_{12} = \frac{n}{2}[2a_1 + (n-1)d]$

$= \frac{12}{2}[2(1000) + 11(260)]$

$= 6(2000 + 2860)$

$= 6(4860)$

$= 29{,}160$ km

Geometric Sequences

A sequence in which a constant r is multiplied by each term to get the next is called a **geometric sequence**. The constant r is called the **common ratio**.

As an example of a geometric sequence, suppose a car loses value or depreciates by 20% a year. The values of a $40,000 car year-by-year for five years are thus $32,000, $25,600, $20,480, $16,384, $13,107. This is not an arithmetic sequence because there is no common difference d. However, multiplying each term by 80%, or 0.8, gives us the next term. So the ratio of each term to the preceding term is 0.8 to 1, and $r = 0.8$.

EXAMPLE

Find the common ratios r of the following geometric sequences:

a. 5, 10, 20, 40, ...,

b. 2, −6, 18, −54, ...

SOLUTION

a. The ratios of the terms are $\frac{10}{5} = 2$, $\frac{20}{10} = 2$, ..., so $r = 2$.

b. The ratios of the terms are $\frac{-6}{2} = -3$, $\frac{18}{-6} = -3$, ..., so $r = -3$.

Just as we had a formula for the nth term of an arithmetic sequence, we have a **formula for the nth term of a geometric sequence**:

$$a_n = a_1 r^{n-1}.$$

Note that the exponent is one less than the number of the term. Remember this formula for the CLEP College Algebra exam. Of course, if n is small, we can find the term if we know the common ratio by just writing out the terms, multiplying each term in turn by this ratio.

EXAMPLE

Find the seventh term of the geometric sequence 2, 6, 18, ...

SOLUTION

$a_1 = 2, r = 3, n = 7$

$a_7 = 2(3)^{7-1}$

$ = 2(3)^6$

$ = 2(729)$

$ = 1458$

Since n is relatively small, we could also have found the solution by just multiplying the terms by 3: 2, 6, 18, 54, 162, 486, 1458.

EXAMPLE
Find the eighth term of the geometric sequence 64, −32, 16, …

SOLUTION

$$a_1 = 64, r = \frac{-1}{2}, n = 8$$

$$a_8 = 64\left(\frac{-1}{2}\right)^{8-1}$$

$$= 64\left(\frac{-1}{2}\right)^{7}$$

$$= 64\left(\frac{-1}{128}\right)$$

$$= \frac{-1}{2}$$

Geometric Series

And just as we had a formula (actually two of them) for the sum of an arithmetic sequence, we also have a formula for the sum of a geometric sequence. We call this a **geometric series**, and the formula requires us to know only the first term a_1 and the common ratio r. The formula is

$$S_n = \frac{a_1\left(1-r^n\right)}{1-r}.$$

Remember this formula for the CLEP College Algebra exam. Again, for a small n, it is often easier to just generate the terms and add them.

EXAMPLE

Find the sum of the first six terms of the geometric series
$3 + 6 + 12 + \ldots$.

SOLUTION

It might be easier just to generate the terms themselves
$(3 + 6 + 12 + 24 + 48 + 96 = 189)$, but by using the formula, we get
(for $a = 3$ and $r = 2$):

$$S_6 = \frac{3(1-2^6)}{1-2}$$
$$= \frac{3(1-64)}{-1}$$
$$= \frac{3(-63)}{-1}$$
$$= 189$$

EXAMPLE

Find the sum of the first five terms of the geometric series
$4 - 20 + 100 + \ldots$.

SOLUTION

Again, it might be easier just to generate the terms themselves
$(4 - 20 + 100 - 500 + 2500 = 2084)$, but by using the formula
($a_1 = 4, r = -5, n = 5$), we get:

$$S_5 = \frac{4\left[1-(-5)^5\right]}{1-(-5)}$$
$$= \frac{4[1-(-3125)]}{6}$$
$$= \frac{4(3126)}{6}$$
$$= 2084$$

Infinite Geometric Series

Even though an infinite series based on an arithmetic sequence will always have an infinite sum, an infinite series based on a geometric sequence can have a finite sum, even though we are adding an infinite number of terms.

If an **infinite geometric series** does have a sum, it is said to be **convergent**. To determine whether a geometric series is convergent, we look at the value of r. If $|r| < 1$, the sum of the infinite geometric series is given by the formula:

$$S_\infty = \frac{a_1}{1-r}.$$

Remember this formula for the CLEP College Algebra exam. It comes from the general formula for a finite geometric series, $S_n = \frac{a_1(1-r^n)}{1-r}$, taking into account that as n gets larger and larger, r^n gets smaller and smaller since r^n is a fraction of a fraction of a fraction, and so on, getting close to (but not reaching) 0. If $|r| \geq 1$, the sum is not convergent—it is divergent and we cannot determine a sum.

EXAMPLE

Find the sum of the geometric series $1 + \frac{1}{2} + \frac{1}{4} + \frac{1}{8} + \frac{1}{16} + \ldots$.

SOLUTION

This is an infinite geometric series with $a_1 = 1, r = \frac{1}{2}$. Since $|r| < 1$, we use the formula $S_\infty = \frac{a_1}{1-r}$:

$$S = \frac{1}{1-\frac{1}{2}}$$

$$S = \frac{1}{\frac{1}{2}}$$

Multiply by $\frac{2}{2}$ to get rid of the fraction $\frac{1}{2}$:

$$S = \frac{1}{\frac{1}{2}}\left(\frac{2}{2}\right) = \left(\frac{2}{1}\right) = 2.$$

EXAMPLE

Find the sum of the geometric series $12 - 3 + \dfrac{3}{4} - \dfrac{3}{16} + \ldots$.

SOLUTION

This is an infinite geometric series with $a_1 = 12, r = -\dfrac{1}{4}$. Since $|r| < 1$, we again use the formula $S_\infty = \dfrac{a_1}{1-r}$:

$$S = \dfrac{12}{1 - \left(\dfrac{-1}{4}\right)}$$

$$S = \left(\dfrac{12}{1 + \dfrac{1}{4}}\right)\left(\dfrac{4}{4}\right)$$

$$S = \dfrac{48}{4+1} = \dfrac{48}{5}$$

EXAMPLE

Find the sum of the geometric series $8 - 12 + 18 - 27 + \ldots$.

SOLUTION

This is an infinite geometric series with $a_1 = 8$, $r = \dfrac{-3}{2}$, but since $|r| = \left|\dfrac{-3}{2}\right| = \dfrac{3}{2} > 1$, this series is divergent. Thus, we cannot determine a sum.

FACTORIALS AND THE BINOMIAL THEOREM

Factorials

The beginning of this chapter addressed the evolution of the various number systems as methods for counting. If we want to count the number of ways a set of objects can be arranged, combined, or chosen, we can do it directly by listing all of the possibilities and counting them, but more efficient ways involve the **counting principle**. The counting principle states that if one event can occur in m ways and a second event can occur in n ways, the total ways both events can occur is $m \times n$.

EXAMPLE

A golfer has four different golf shirts and three different pair of golf pants. How many outfits are possible?

SOLUTION

There are 4(3) = 12 different outfits possible.

The counting principle also works when there are more than two items being considered.

EXAMPLE

A pizza shop sells small, medium, and large pizzas. Each pizza can be ordered in thin or thick crust. And each pizza has a choice of five toppings—extra cheese, pepperoni, sausage, mushrooms, or anchovies. If a one-topping pizza is ordered, how many different pizzas choices are there?

SOLUTION

There are 3(2)(5) = 30 different pizza choices.

EXAMPLE

A lottery involves a three-digit number. How many number choices are there?

SOLUTION

There are 9 possible first digits (1 — 9, since if 0 were allowed in the first position, it would be a two-digit number), 10 possible second digits, and 10 possible third digits. So there are 9(10)(10) = 900 three-digit numbers. Note that in this example, we can use a digit over again, so 444 could be one of the 900 three-digit numbers.

Similarly, using the counting principle, we can determine, for example, how many ways five people can line up. The first position can be any of the five people, but as we get to the next position, only four, three, and two people are left, until there is only one choice (last person standing) for the last position. So there are $5 \times 4 \times 3 \times 2 \times 1 = 120$ ways.

CHAPTER 6: NUMBER SYSTEMS AND OPERATIONS

A special symbol for this calculation is $n!$, called **n factorial**, and it is written as $n! = n \times (n-1) \times (n-2) \times \ldots \times 3 \times 2 \times 1$. It is the product of n and every number less than n all the way down to 1. Note that $0!$ is defined as 1, or $0! = 1$. So five people can line up in $5! = 120$ ways.

EXAMPLE

Evaluate $\dfrac{n!}{(n+2)!}$

SOLUTION

$\dfrac{n \times (n-1) \times (n-2) \times \ldots \times 1}{(n+2) \times (n+1) \times n \times (n-1) \times (n-2) \times \ldots \times 1} = \dfrac{1}{(n+2) \times (n+1)}$. Note that all the terms $n \times (n-1) \times \ldots \times 1$ in the numerator and the denominator cancel each other out.

Factorials are handy when choosing r objects out of n objects, either when order makes a difference (called a **permutation** and denoted $_nP_r = \dfrac{n!}{(n-r)!}$) or when order doesn't make a difference (called a **combination** and denoted $_nC_r = \dfrac{n!}{r!(n-r)!}$). For the CLEP College Algebra exam, it is important to know how to calculate such factorials. Fortunately, many factors cancel each other out in permutations and combinations, so the arithmetic is usually simple. For example,

$$_{12}P_2 = \dfrac{12!}{10!} = \dfrac{12 \times 11 \times \cancel{10} \times \cancel{9} \times \cancel{8} \times \cancel{7} \times \cancel{6} \times \cancel{5} \times \cancel{4} \times \cancel{3} \times \cancel{2} \times \cancel{1}}{\cancel{10} \times \cancel{9} \times \cancel{8} \times \cancel{7} \times \cancel{6} \times \cancel{5} \times \cancel{4} \times \cancel{3} \times \cancel{2} \times \cancel{1}}$$
$$= 12 \times 11 = 132.$$

Likewise,

$$_{12}C_2 = \dfrac{12!}{2!\,10!} = \dfrac{12 \times 11 \times \cancel{10} \times \cancel{9} \times \cancel{8} \times \cancel{7} \times \cancel{6} \times \cancel{5} \times \cancel{4} \times \cancel{3} \times \cancel{2} \times \cancel{1}}{2 \times 1 \times \cancel{10} \times \cancel{9} \times \cancel{8} \times \cancel{7} \times \cancel{6} \times \cancel{5} \times \cancel{4} \times \cancel{3} \times \cancel{2} \times \cancel{1}}$$
$$= \dfrac{12 \times 11}{2 \times 1} = 66.$$

Permutations and combinations are always whole numbers, which means every number in the denominator always cancels out. Also, since it is more common in algebra to use parentheses rather than × to indicate multiplication, the rest of the calculations in this chapter will use parentheses.

EXAMPLE

Find the value of $\dfrac{10!}{8!}$.

SOLUTION

$$\frac{10!}{8!} = \frac{10(9)(8)(7)\ldots(3)(2)(1)}{(8)(7)\ldots(3)(2)(1)}$$

Since all these numbers are multiplied, we can cancel all numbers from 8 down to 1 in the numerator and the denominator.

$$\frac{10(9)(\cancel{8})(\cancel{7})\ldots(\cancel{3})(\cancel{2})(\cancel{1})}{(\cancel{8})(\cancel{7})\ldots(\cancel{3})(\cancel{2})(\cancel{1})} = 10(9) = 90.$$

We recognize that $\dfrac{10!}{8!} = \dfrac{10(9)(8!)}{(8!)}$ and that the 8! cancels, so to save time, we don't have to write all the factors $(8)(7)(6)\ldots(3)(2)(1)$ in the numerator and the denominator only to then cross them out.

EXAMPLE

The letters of the set $\{A, B, C, D, E, F, G\}$ can be arranged to form ordered codes of three letters in $_7P_3 = \dfrac{7!}{(7-3)!}$ ways. Evaluate $_7P_3$.

SOLUTION

$$_7P_3 = \frac{7!}{(7-3)!} = \frac{7!}{4!} = \frac{7(6)(5)(4!)}{(4!)} = 7(6)(5) = 210.$$

Here, we recognized that the 4! cancels out of the numerator and the denominator.

CHAPTER 6: NUMBER SYSTEMS AND OPERATIONS

EXAMPLE

An office has 16 workers. The boss decides to choose a manager and assistant from that group. This can be done in $_{16}P_2 = \dfrac{16!}{(16-2)!}$ ways. What is the value of $_{16}P_2$?

SOLUTION

$$_{16}P_2 = \dfrac{16!}{(16-2)!} = \dfrac{16!}{14!} = \dfrac{16(15)(14!)}{(14!)} = 16(15) = 240.$$

Note: This is a permutation of 16 things choosing 2 because order is important. John as the manager and Jane as the assistant is different from Jane as the manager and John as the assistant.

EXAMPLE

Find the value of $\dfrac{8!}{3!5!}$.

SOLUTION

$$\dfrac{8!}{3!5!} = \dfrac{(8)(7)(6)(5!)}{(3)(2)(1)(5!)} = \dfrac{(8)(7)(6)}{(3)(2)(1)}$$

The 6 in the numerator cancels out the 3 and the 2 in the denominator, so we have $8(7) = 56$.

We handle the $\dfrac{8!}{5!}$ part of the calculation as we did in the examples above, by writing the symbol 5! in the numerator and in the denominator and canceling them out.

EXAMPLE

An ice cream shop offers 20 flavors. Derek purchases a dish of three different flavors. He can choose the flavors in $_{20}C_3 = \dfrac{20!}{3!(20-3)!}$ ways. How many choices does Derek have?

SOLUTION

$$_{20}C_3 = \dfrac{20!}{3!(20-3)!} = \dfrac{20!}{3!\,17!} = \dfrac{(20)(19)(18)(17!)}{(3)(2)(1)(17!)} = \dfrac{(20)(19)(18)}{(3)(2)(1)} = 1,140.$$

Note: This is a combination because order is not important. A dish of chocolate, vanilla, and strawberry is the same as a dish of strawberry, vanilla, and chocolate.

EXAMPLE

An office has 16 workers. The boss decides to choose two managers from that group. This can this be done in $_{16}C_2 = \dfrac{16!}{2!(16-2)!}$ ways. Calculate $_{16}C_2$.

SOLUTION

$$_{16}C_2 = \dfrac{16!}{2!(16-2)!} = \dfrac{16!}{2!\,14!} = \dfrac{(16)(15)(14!)}{(2)(1)(14!)} = \dfrac{16(15)}{(2)(1)} = 120.$$

Note: Since both people chosen have the same rank (as opposed to manager and assistant in a prior example), order is not important. Choosing John and Jane as managers is the same thing as choosing Jane and John.

Binomial Theorem

Let's go back to some basic algebra. Chapter 3 worked with expanding binomials. By using multiplying, we find that

$$(a \pm b)^2 = a^2 \pm 2ab + b^2$$
$$(a \pm b)^3 = a^3 \pm 3a^2b + 3ab^2 \pm b^3$$
$$(a \pm b)^4 = a^4 \pm 4a^3b + 6a^2b^2 \pm 4ab^3 + b^4$$

Note: We use $(a \pm b)^n$ here rather than $(x \pm y)^n$ to avoid confusion. If we were finding powers of $(2x + 3y)$, for example, it is less confusing to substitute $a = 2x$ and $b = 3y$ than to say $x = 2x$ and $y = 3y$. Just remember that we are actually using a to mean the first term and b to mean the second term, whatever they are.

We did not expand expressions with higher powers in Chapter 3 because of all the work involved. However, the **binomial theorem** gives us a quick and easy way to do this binomial expansion without having to go through all the work. The binomial theorem states:

$$(a+b)^n = {}_nC_0 a^n + {}_nC_1 a^{n-1}b + {}_nC_2 a^{n-2}b^2 + {}_nC_3 a^{n-3}b^3 + \ldots + {}_nC_{n-2} a^2 b^{n-2}$$
$$+ {}_nC_{n-1} ab^{n-1} + {}_nC_n b^n$$

Quick and easy? Yes, if we find the terms by the following procedures, which use a quick and easy shortcut for calculating ${}_nC_r = \dfrac{n!}{r!(n-r)!}$ (or an alternative quick and easy way to find ${}_nC_r$ by using Pascal's triangle), combined with a quick and easy way to find the exponents of the variables.

Exponents of the Variables

Let's first examine the powers in the expansion of $(a+b)^2$. The exponent of the first term, a, goes down from 2 to 1 to 0 (there is no a in the last term), and the exponent of the last term, b, goes up from 0 (there is no variable b in the first term) to 1 to 2. Observe that the sum of the exponents in each term is 2, the power of the binomial, and that the number of terms in the expansion is 3, one more than the power.

Next, in $(a+b)^3$, the a variable's exponent goes down: 3, 2, 1, 0, while the b variable's exponent goes up: 0, 1, 2, 3. Note also that the sum of the exponents of the two variables in each term is always 3, the power of the binomial, and here we have $3 + 1 = 4$ terms.

We are beginning to see a pattern here. In $(a+b)^4$, the a variable's exponent goes down: 4, 3, 2, 1, 0, while the b variable's exponent goes up: 0, 1, 2, 3, 4. Note also that the sum of the exponents of the two variables in each term is always 4, the power of the binomial, and the number of terms is $4 + 1 = 5$.

So we can rather quickly (and easily) write the variables for any term in the expansion, just knowing the power, n, and the number of terms we want, r.

Coefficients of the Binomial Expression, ${}_nC_r$

A value for ${}_nC_r$ can be found quickly and easily if we do all of the cancellations mentally, writing only the minimal amount of calculation for ${}_nC_r = \dfrac{n!}{r!(n-r)!}$. We saw in the last example that ${}_{16}C_2 = \dfrac{16!}{2!(16-2)!} = \dfrac{16!}{2!\,14!} = \dfrac{16(15)(14!)}{2(1)(14!)}$ and the

14! cancels out. The first thing we want to do mentally is to cancel out the 14!, so we can quickly write $_{16}C_2 = \frac{(16)(15)}{(2)(1)} = 8(15) = 120$. Note that the denominator will *always* cancel into the numerator.

To make the calculation of each $_nC_r$ easy, we must recognize that $\frac{16!}{2!\,14!}$ is the same as $\frac{16!}{14!\,2!}$, which means we should choose the larger factorial as the one to cancel out. We are then left (after canceling the 14! in this example) with $\frac{(16)(15)}{(2)(1)}$. We now see another fact that will streamline the calculation even further: The number of factors in the numerator always matches the number of factors in the denominator.

Thus, we can look at a seemingly complicated calculation (especially if we write it all out), such as $_{29}C_3$ and immediately write $_{29}C_3 = \frac{(29)(28)(27)}{(3)(2)(1)} = (29)(14)(9) = 3654$. Note that there are only three factors in both the numerator and the denominator.

In summary, for any combination, $_nC_r$ reduces to writing only as many factors as the smaller number, r or $(n - r)$, in the numerator and in the denominator. After the cancellations, the whole thing reduces to a simple multiplication problem that can be done quickly by hand or with a calculator.

Pascal's Triangle

Another quick and easy way to find the coefficients in a binomial expansion (particularly for $n \leq 6$) is to use **Pascal's triangle**. Pascal's triangle isn't included by name in the CLEP College Algebra test description, but it is a handy shortcut for finding coefficients in the binomial expansion, especially for smaller exponents, as shown in the following examples. Pascal's triangle for exponents (rows) up to 6 is:

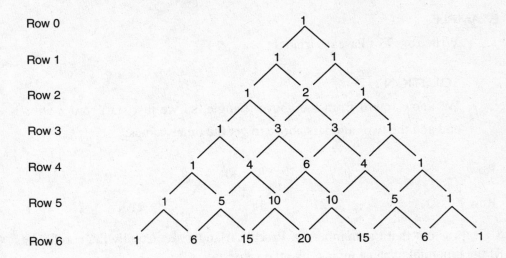

The array of numbers in Pascal's triangle is created by a series of rows with the number 1 on the outside. Any element is found by adding the elements on the left and the right from the row above it. The first seven rows of Pascal's triangle are shown here. Note that the first row is called row 0 and has only one element, 1. The remaining row numbers are n, which, as we said, is the exponent of the binomial we are expanding. All outside values (left and right) in each row are 1's, and the second term and next to last term in each row are the same as the row number. Also note that Pascal's triangle has symmetry about the center.

The only memorization needed for Pascal's triangle is that the first two rows are

Row 0   1

Row 1   1   1

All the other numbers can be generated just by adding the two numbers above each of them.

EXAMPLE

Write row 7 of Pascal's triangle.

SOLUTION

We know row 6 from the above triangle, so we just start row 7 with 1 and add the two numbers above to get the other values:

Row 6	1	6	15	20	15	6	1
Row 7	1	7	21	35	35	21	7	1

It turns out that the numbers in Pascal's triangle are actually the coefficients of the binomial expansion, just like the values of $_nC_r$.

If the sign in the binomial is positive, all coefficients are positive. If the sign in the binomial is negative, the sign of the coefficient starts positive and then alternates.

Steps for Finding the Binomial Expansion

So now we can put it all together and list the steps for the binomial expansion.

Step 1: Determine the number of terms (the power of the binomial plus 1), and the signs of the terms:

$(a + b)^n = \underline{} + \underline{} + \underline{} + \underline{} + \ldots$ ($n + 1$ terms).

$(a - b)^n = \underline{} - \underline{} + \underline{} - \underline{} + \ldots$ ($n + 1$ terms).

Step 2: Fill in the first variables, starting with a^n down to a^0 ($= 1$).

$(a + b)^n = \underline{a^n} + \underline{a^{n-1}} + \underline{a^{n-2}} + \underline{a^{n-3}} + \ldots + \underline{a^1} + \underline{a^0}$.

Step 3: Fill in the second variables, starting with b^0 ($= 1$) up to b^n.

$(a + b)^n = \underline{a^n} + \underline{a^{n-1}b} + \underline{a^{n-2}b^2} + \underline{a^{n-3}b^3} + \ldots + \underline{ab^{n-1}} + \underline{b^n}$.

Note that the exponents in each term add up to n.

Step 4: Fill in the coefficients by using $_nC_r$ or Pascal's triangle (using row n).

To find $(x + y)^5$ or $(x - y)^5$, the coefficients from row 5 of Pascal's triangle are 1 5 10 10 5 1, so we can now quickly expand $(x + y)^5 = x^5 + 5x^4y + 10x^3y^2 + 10x^2y^3 + 5xy^4 + y^5$.

Likewise, $(x - y)^5 = x^5 - 5x^4y + 10x^3y^2 - 10x^2y^3 + 5xy^4 - y^5$; since the binomial has a minus sign, we start positive, then alternate signs.

Of course, by using the binomial theorem, $(x + y)^5 = {}_5C_0 x^5 + {}_5C_1 x^4 y + {}_5C_2 x^3 y^2 + {}_5C_3 x^2 y^3 + {}_5C_1 xy^4 + {}_5C_5 y^5$, we would get the same result.

EXAMPLE

Use Pascal's triangle to expand $(x + y)^7$ and $(x - y)^7$.

SOLUTION

Here $a = x$ and $b = y$. Row 7 of Pascal's triangle was constructed in an exercise above as 1 7 21 35 35 21 7 1. We know there will be eight terms. We can fill in the signed coefficients for each term, and then the variables for the other terms according to steps 2 and 3 above. The first x factor is $(1) x^7 y^0 = x^7$, and the exponent of x reduces by 1 in each succeeding term. The first y factor is y^1, starting in the second term since the first term has $y^0 = 1$. It is straightforward to fill in the eight terms (inserting the coefficients from Pascal's triangle) to get:

$(x + y)^7 = x^7 + 7x^6y + 21x^5y^2 + 35x^4y^3 + 35x^3y^4 + 21x^2y^5 + 7xy^6 + y^7$

$(x - y)^7 = x^7 - 7x^6y + 21x^5y^2 - 35x^4y^3 + 35x^3y^4 - 21x^2y^5 + 7xy^6 - y^7$

EXAMPLE

Expand $(x + 7)^4$.

SOLUTION

Here $a = x$ and $b = 7$, and the method is the same as in the last example.

$(x + 7)^4 = x^4 + 4x^3(7) + 6x^2(7)^2 + 4x(7)^3 + (7)^4$

$= x^4 + 28x^3 + 294x^2 + 1372x + 2401$

EXAMPLE

Expand $(m^2 - 2n)^3$.

SOLUTION

Here, $a = m$ and $b = 2n$. Since the sign in the binomial is minus, the signs of the terms will begin positive and alternate. We can immediately write

$$(m^2 - 2n)^3 = (1)(m^2)^3 - 3(m^2)^2(2n) + 3(m^2)^1(2n)^2 - (1)(2n)^3$$
$$= m^6 - 3m^4(2n) + 3m^2(4n^2) - 8n^3$$
$$= m^6 - 6m^4n + 12m^2n^2 - 8n^3$$

It is highly unlikely that the CLEP College Algebra exam will ask you to completely expand a binomial because there are just too many terms. Knowledge of the binomial theorem and Pascal's triangle can come in handy to find just one particular term, however, even if the binomial is raised to a high power.

We can use the following generalization to find the rth term of $(a + b)^n$.

1. The coefficient of the term is the rth term of row n of Pascal's triangle, or can be calculated from $_nC_{r-1}$.

2. The b variable is b^{r-1}.

3. The a variable is a^{n-r+1}, or more easily, the exponent of the a variable is n minus the exponent of b (find the b variable first).

4. If the binomial is $(a - b)^n$ and r is odd, the term is positive; if r is even, the term is negative.

EXAMPLE

Find the fifth term of $(x + y)^{11}$.

SOLUTION

Here, $n = 11$, $r = 5$, $a = x$, and $b = y$.

Using the binomial formula and $_nC_{r-1}$, we get

$$\begin{aligned}
{}_{11}C_{5-1}x^{11-5+1}y^{5-1} &= {}_{11}C_4 x^7 y^4 \\
&= \frac{11!}{7!\,4!} x^7 y^4 \\
&= \frac{(11)(10)(9)(8)}{(4)(3)(2)(1)} x^7 y^4 \\
&= 330 x^7 y^4
\end{aligned}$$

To find the coefficient of the fifth term of the 11th row of Pascal's triangle, we don't have to write out every value, only the first five in each row (see below).

Row 0						1				
Row 1							1	1		
Row 2						1	2	1		
Row 3					1	3	3	1		
Row 4				1	4	6	4	1		
Row 5				1	5	10	10	5		
Row 6			1	6	15	20	15			
Row 7			1	7	21	35	35			
Row 8		1	8	28	56	70				
Row 9		1	9	36	84	126				
Row 10	1	10	45	120	210					
Row 11	1	11	55	165	330					

Putting this all together, we also get that the fifth term of $(x+y)^{11}$ is $330\, x^7 y^4$.

EXAMPLE

Find the 19th term of $(a-b)^{19}$.

SOLUTION

Here, $n = 19$, and $r = 19$. The sign in the binomial is minus, and $r = 19$, which is odd, so the 19th term is positive. The coefficients of the 2nd and 19th terms in Pascal's triangle are 19, the same as the row number, as explained above. Therefore, we get right away that the answer is $19ab^{18}$.

The same answer can be found by using $_nC_{r-1}$ for the coefficients:

$$_{19}C_{18}a^1b^{18}$$
$$= \frac{19!}{1!\,18!}ab^{18}$$
$$= 19ab^{18}$$

By now it is obvious that all the cancellations when using $_nC_r$ make this method easy when r is close to 1 or n.

EXAMPLE

Find the fourth term of $\left(x^3 - \frac{1}{2}y\right)^8$.

SOLUTION

For this example, $a = x^3$, $b = \frac{1}{2}y$, $n = 8$, and $r = 4$, which is even, so the term is negative. First, we find the coefficient. Using the binomial theorem, we get $_8C_3 = \frac{8!}{5!\,3!} = \frac{(8)(7)(6)}{(3)(2)(1)} = (8)(7) = 56.$

The second factor in the binomial is $\left(\frac{1}{2}y\right)$, and it is raised to the third power, since $r - 1 = (4 - 1) = 3$.

The first factor will be raised to the fifth power (the power of the second factor in the binomial subtracted from the power of the binomial, or $(8 - 3) = 5$).

Therefore, putting together all of the steps, the fourth term in the binomial expansion is $56(x^3)^5\left(\frac{1}{2}y\right)^3 = 56x^{15}\left(\frac{1}{8}y^3\right) = 7x^{15}y^3.$

Using Pascal's triangle to find the coefficient would take more time, but if we reuse the triangle we constructed above for 11 rows, we see that the fourth term in row 8 is also 56.

MATRICES

Addition, Subtraction, and Scalar Multiplication

A **matrix** is a rectangular array of numbers. **Matrices** (plural) are made up of rows and columns. A **square matrix** has the same number of rows and columns. We ordinarily write brackets around matrices. The **dimension** of a matrix is given as "number of rows × number of columns." Capital letters are usually used to denote matrices. Each value in a matrix is referred to as an **element**, sometimes written as a_{ij}, where i is the row and j is the column.

EXAMPLE

Find the dimensions of each of the following matrices:

a. $\begin{bmatrix} 2 & 5 & -3 \\ 4 & 1 & 5 \\ -2 & 0 & 7 \end{bmatrix}$ b. $[\,5\ 2\,]$ c. $\begin{bmatrix} -3 \\ \pi \\ 0 \end{bmatrix}$

SOLUTION

a. 3 × 3

b. 1 × 2

c. 3 × 1

In this chapter, as on the CLEP College Algebra test, we deal with 2 × 2 matrices, although as long as two or more matrices have exactly the same dimensions, addition and subtraction holds for all sizes of matrices. Scalar multiplication also holds for all sizes of matrices, but there are restrictions on multiplying two matrices, a topic not covered in the CLEP College Algebra test, and therefore beyond the scope of this book.

EXAMPLE

If $M = \begin{bmatrix} 8 & 4 \\ -1 & 5 \end{bmatrix}$ and $N = \begin{bmatrix} -2 & 4 \\ 3 & -1 \end{bmatrix}$, find

a. $M + N$
b. $M - N$

SOLUTION

Since both matrices have dimensions 2×2, we can add and subtract them, element by corresponding element.

a. $M + N = \begin{bmatrix} 8-2 & 4+4 \\ -1+3 & 5-1 \end{bmatrix} = \begin{bmatrix} 6 & 8 \\ 2 & 4 \end{bmatrix}$

b. $M - N = \begin{bmatrix} 8+2 & 4-4 \\ -1-3 & 5+1 \end{bmatrix} = \begin{bmatrix} 10 & 0 \\ -4 & 6 \end{bmatrix}$

Scalar multiplication of matrices involves multiplying a matrix by a scalar k (a real number). The result is another matrix with the same dimensions as the original one with every element of A multiplied by k.

EXAMPLE

Multiply $-4 \begin{bmatrix} 5 & \pi \\ -2 & -3x \end{bmatrix}$.

SOLUTION

$\begin{bmatrix} -20 & -4\pi \\ 8 & 12x \end{bmatrix}$

EXAMPLE

Multiply $6 \begin{bmatrix} y & 10 \\ -\dfrac{2}{3} & \dfrac{x^2}{2} \end{bmatrix}$.

SOLUTION

$\begin{bmatrix} 6y & 60 \\ -4 & 3x^2 \end{bmatrix}$

Determinants

Every square matrix is associated with a number called the **determinant**. The determinant of a matrix A is denoted $|A|$. The determinant of a 2×2 matrix is defined as follows:

The determinant of the matrix $\begin{bmatrix} a & b \\ c & d \end{bmatrix}$ is written with straight brackets as $\begin{vmatrix} a & b \\ c & d \end{vmatrix}$. Its value is $ad - bc$. To remember which elements get multiplied, we use a visual of writing a capital X, starting at the top left. The first stroke ($\diagdown$) is the first multiplication, and the second stroke ($\diagup$) is the multiplication that gets subtracted from the first. Thus, we get $ad - bc$.

EXAMPLE

Find the values of the following determinants:

a. $\begin{vmatrix} 7 & -2 \\ -4 & 5 \end{vmatrix}$

b. $\begin{vmatrix} -5 & 1 \\ 0 & -4 \end{vmatrix}$

c. $\begin{vmatrix} \dfrac{-1}{2} & \dfrac{-2}{3} \\ \dfrac{3}{2} & 6 \end{vmatrix}$

d. $\begin{vmatrix} 1 & 2 & 3 \\ -1 & -2 & -3 \end{vmatrix}$

SOLUTION

a. $7(5) - (-2)(-4) = 35 - 8 = 27.$

b. $-5(-4) - (1)(0) = 20 - 0 = 20.$

c. $-\dfrac{1}{2}(6) - \left(\dfrac{-2}{3}\right)\left(\dfrac{3}{2}\right) = -3 + 1 = -2.$

d. This is not a square matrix, so it has no determinant.

Chapter 5 covered three techniques to solve a system of two equations with two variables: graphing, substitution, and elimination. A fourth method, called Cramer's Rule, uses determinants to solve such a system of equations. If a problem on the CLEP College Algebra test uses Cramer's Rule, the formula will be given, but we must be able to calculate the value of the determinants to find one or both of the variables.

EXAMPLE

The solution to the system of equations: $\begin{cases} 4x+3y=14 \\ 2x-5y=20 \end{cases}$ given by Cramer's rule is $x = \dfrac{\begin{vmatrix} 14 & 3 \\ 20 & -5 \end{vmatrix}}{\begin{vmatrix} 4 & 3 \\ 2 & -5 \end{vmatrix}}$ and $y = \dfrac{\begin{vmatrix} 4 & 14 \\ 2 & 20 \end{vmatrix}}{\begin{vmatrix} 4 & 3 \\ 2 & -5 \end{vmatrix}}$. Find the solution to the system.

SOLUTION

$$x = \frac{14(-5)-3(20)}{4(-5)-3(2)}$$
$$= \frac{-70-60}{-20-6}$$
$$= \frac{-130}{-26}$$
$$= 5$$

$$y = \frac{4(20)-14(2)}{4(-5)-3(2)}$$
$$= \frac{80-28}{-20-6}$$
$$= \frac{52}{-26}$$
$$= -2$$

So the solution is $x = 5, y = 2$.

CHAPTER 7

Attacking the CLEP College Algebra Exam

CHAPTER 7

ATTACKING THE CLEP COLLEGE ALGEBRA EXAM

As described in Chapter 1, the CLEP College Algebra exam consists of approximately 60 questions, mostly multiple-choice, and you will have 90 minutes to complete the exam.

The exam covers material that is typically part of a one-semester college course. The exam places little emphasis on arithmetic calculations, and it does not contain any questions that require the use of a graphing calculator, even though an online scientific calculator is available to test-takers. If you are unsure of your arithmetic at any point, use it.

The chapters of this book are modeled on the breakdown of topics found on the CLEP College Algebra exam. The approximate breakdown and number of problems you can expect are as follows:

Algebraic Operations: 25%, or about 15 problems (Chapter 3)

Functions and Their Properties: 25%, or about 15 problems (Chapter 4)

Equations and Inequalities: 30%, or about 18 problems (Chapter 5)

Number Systems and Operations: 20%, or about 12 problems (Chapter 6)

CLASSIFYING CLEP PROBLEMS

The two types of problems on the CLEP College Algebra exam are classified as routine problems and nonroutine problems.

1. **Routine problems.** These problems specifically ask you to perform a task. It is clear what the problem asks you to do and there is little creativity or ingenuity involved in the solution. Approximately 50% (30 problems) will be of this routine nature.

   Routine questions are straightforward. For example, a child first learning arithmetic might be asked to do the subtraction $8 - 5$. If the child knows the technique of subtraction, it becomes routine and the child has no problem.

2. **Nonroutine problems.** These problems require an understanding of concepts and the application of skills. With nonroutine problems, knowing how to do the task is not enough. You must know what task is called for and how to apply it to the problem. Approximately 50% (30 problems) will be of the nonroutine nature.

   Nonroutine questions are not straightforward. You are not told specifically what to do, which, for most students, makes these questions appreciably more difficult. For example, a child who is asked how many pieces of candy she will have if she has 8 pieces of candy and gives away 5 of them needs to not only know how to do subtraction (routine), but also that it is *necessary* to do subtraction (nonroutine).

For your benefit, in the answer keys for the two practice exams included with this book, the solutions indicate whether the problem is routine or nonroutine. You can approach the solution either directly or by trial-and-error. Use the strategy that is more comfortable for you. In the direct approach, you are given a problem and a task to do. The direct approach requires that you do the task and determine which of the five answer choices fit. If the CLEP College Algebra exam required you to show your work, you would have to do all the problems by the direct approach. Often there are several direct ways to solve a problem, all of them correct. The following problems show some alternatives. The trial-and-error approach tests each of the answer choices to see which fits the criteria of the problem. There will be only one solution that "works."

For example, suppose you were asked to completely factor $-2x^3 - 6x^2 + 36x$ and the five answer choices were:

(A) $-2x(x^2 - 3x - 18)$

(B) $-2x(x^2 + 3x - 18)$

(C) $-2x(x - 6)(x + 3)$

(D) $-2x(x + 6)(x - 3)$

(E) $(-2x - 6)(x + 6)$

This is a routine problem. There is a specific task you are asked to do: complete factorization.

By the direct approach, you would perform these steps:

$-2x^3 - 6x^2 + 36x$

$-2x(x^2 + 3x - 18)$

$-2x(x + 6)(x - 3)$

The correct answer is (D).

By the trial-and-error approach for the same problem, you would multiply out each of the five choices until you get the correct answer. However, eliminating any of the choices before you do this shortens the time to complete the problem. In this case, the words "completely factor" are the clue. You should recognize that answer choices (A) and (B) may not be completely factored, since they contain quadratics. Therefore, you can save time by starting your trial-and-error with choice (C).

So, for choices (C) through (E),

(C) $-2x(x - 6)(x + 3) = -2x(x^2 - 3x - 18) = -2x^3 + 6x^2 + 36x$. This is not the original expression, so this isn't the answer.

(D) $-2x(x + 6)(x - 3) = -2x(x^2 + 3x - 18) = -2x^3 - 6x^2 + 36x$. Bingo! This is the answer, and you don't even have to work with (E). Just go on to the next problem.

The next section presents some challenging problems from each of the four main topics on the CLEP College Algebra exam and explains techniques to solve them.

SAMPLE PROBLEMS

Algebraic Operations

1. Find the value of $\left(\dfrac{-2x^2}{3y^{-3}}\right)^{-3}$.

    (A) $\dfrac{-2y^9}{3x^6}$

    (B) $\dfrac{-8y^9}{27x^6}$

    (C) $\dfrac{-27y^6}{8x^6}$

    (D) $\dfrac{-27}{8x^6 y^9}$

    (E) $\dfrac{-27x^6 y^9}{8}$

    SOLUTION

    This is a routine problem. You need to know how to raise expressions to powers as well as the definition of negative exponents. You can directly solve this problem in two ways. By raising to powers first and then using the definition of negative powers, the solution is

    $$\left(\dfrac{-2x^2}{3y^{-3}}\right)^{-3} = \dfrac{(-2)^{-3} x^{-6}}{3^{-3} y^9} = \dfrac{3^3}{(-2)^3 x^6 y^9} = \dfrac{-27}{8x^6 y^9}.$$

    By using the definition of negative powers first and then raising to powers, the solution is $\left(\dfrac{-2x}{3y^{-3}}\right)^{-3} = \left(\dfrac{3y^{-3}}{-2x^2}\right)^3 = \left(\dfrac{3}{-2x^2 y^3}\right)^3 = \dfrac{-27}{8x^6 y^9}.$

    The correct answer is (D).

2. What is the difference between the multiplicative inverse of a and the additive inverse of a?

(A) $a - \dfrac{1}{a}$

(B) $\dfrac{1}{a^2}$

(C) $-\dfrac{1}{a^2}$

(D) $\dfrac{1-a^2}{a}$

(E) $\dfrac{1+a^2}{a}$

SOLUTION

This is a nonroutine problem. First, you need to know what the additive and multiplicative inverses are. Second, you must know that since you are asked to find the difference, you must subtract them. And finally, you must know how to get a common denominator. So a combination of skills is required to solve the problem.

Multiplicative inverse: $\dfrac{1}{a}$. Additive inverse: $-a$

Difference: $\dfrac{1}{a} - (-a) = \dfrac{1}{a} + a\left(\dfrac{a}{a}\right) = \dfrac{1+a^2}{a}$.

The correct answer is (E).

You could also use the trial-and-error technique, working numerically rather than algebraically. Choose a small value for a, say, $a = 2$. The multiplicative inverse is $\dfrac{1}{2}$ and the additive inverse is -2. So $\dfrac{1}{2} - (-2) = \dfrac{1}{2} + 2 = \dfrac{5}{2}$. Now plug in $a = 2$ to all of the choices. Only choice (E) has a value of $\dfrac{5}{2}$. This approach is a little time-consuming, but it works.

3. Simplify $\dfrac{\dfrac{1}{x}+y}{\dfrac{1}{y}+x}$.

(A) xy

(B) $\dfrac{y}{x}$

(C) $\dfrac{x}{y}$

(D) $1+xy$

(E) $x+xy$

SOLUTION

This is a routine problem of simplifying complex fractions by multiplying both numerator and denominator by the common denominator.

$$\left(\dfrac{\dfrac{1}{x}+y}{\dfrac{1}{y}+x}\right)\left(\dfrac{xy}{xy}\right)=\dfrac{y+xy^2}{x+x^2y}=\dfrac{y(1+xy)}{x(1+xy)}=\dfrac{y}{x}.$$

Another method would be to convert the numerator and denominator into fractions and then do the division.

$$\left(\dfrac{\dfrac{1}{x}+y}{\dfrac{1}{y}+x}\right)=\dfrac{\left(\dfrac{1+xy}{x}\right)}{\left(\dfrac{1+xy}{y}\right)},$$ which is the division problem $\dfrac{1+xy}{x}\div\dfrac{1+xy}{y}.$

This becomes (by the "invert and multiply" rule for division of fractions)
$$\dfrac{1+xy}{x}\cdot\dfrac{y}{1+xy}=\dfrac{y}{x}.$$

The correct answer is (B).

CHAPTER 7: ATTACKING THE CLEP COLLEGE ALGEBRA EXAM

4. If x is an integer and $-1 < \sqrt{x+2} < 3$, how many solutions does this inequality have?

(A) 5

(B) 6

(C) 7

(D) 8

(E) 9

SOLUTION

If you were asked to simply solve the inequality, this would be a routine problem. The twist is the added requirement that x be an integer. The problem requires you to know how to solve inequalities as well as to know what an integer is.

Write the inequality as two inequalities, and then combine your answers; this is the routine part.

$-1 < \sqrt{x+2}$ $\sqrt{x+2} < 3$

$1 < x+2$ $x+2 < 9$

$-1 < x$ $x < 7$

So $-1 < x < 7$.

Check to make sure none of these values for x make the radicand $(x+2)$ negative (they don't).

Now for the nonroutine part of the problem. Seven integers satisfy the requirement: 0, 1, 2, 3, 4, 5, 6. The correct answer is (C).

Functions and Their Properties

5. If $f(x)$ is given in the table below, find $f(g(2) + 2)$

x	2	4	6	8	10
$f(x)$	8	2	10	2	4
$g(x)$	4	8	2	10	6

(A) 2

(B) 4

(C) 6

(D) 8

(E) 10

SOLUTION

This is a mostly routine problem requiring you to understand composition of functions. Once you find $g(2)$, you must add 2 to it before using the f function.

$f(g(2) + 2) = f(4 + 2) = f(6) = 10.$

The correct answer is (E).

6. What is the domain of $y = \sqrt{x+1} + \sqrt{x} + \sqrt{x-1}$?

(A) $x \geq 1$

(B) $x \leq 1$

(C) $x \leq 0$

(D) $x \geq -1$

(E) $x \leq -1$

SOLUTION

This is a nonroutine problem. You must know that the domain of a sum of radicals is the domain that satisfies *every* radical in the sum and that radicands cannot be negative, the domain of $y = \sqrt{x+1}$ is $x \geq -1$. The domain of $y = \sqrt{x}$ is $x \geq 0$. The domain of $y = \sqrt{x-1}$ is $x \geq 1$. Since x

must be a number in the domain of all three square roots, $x \geq 1$. The correct answer is (A).

Choices can be eliminated by the trial-and-error approach by substituting numbers into the function. For instance, choosing $x = 0$ would eliminate choice (C), and choosing $x = -1$ would eliminate choices (D) and (E).

Note: Chapter 6 discusses imaginary numbers. When domain is requested, however, assume real numbers, not imaginary numbers.

7. Find the equation of the line passing through the points $(5, -3)$ and $(-1, 1)$.

(A) $2x + 3y = -1$

(B) $2x + 3y = 1$

(C) $2x + 3y = 5$

(D) $2x - 3y = 1$

(E) $2x - 3y = -1$

SOLUTION

Finding the equation of a line given two points is a routine problem. First, find the slope of the line through the two points:

$$m = \frac{y_2 - y_1}{x_2 - x_1} = \frac{1+3}{-1-5} = \frac{4}{-6} = -\frac{2}{3}.$$

Then write the equation of the line by using either point.

$$y - y_2 = m(x - x_2) \Rightarrow y - 1 = -\frac{2}{3}(x + 1) \Rightarrow 3y - 3 = -2x - 2,$$

which in general form is $2x + 3y = 1$.

$$y - y_1 = m(x - x_1) \Rightarrow \text{or } y + 3 = -\frac{2}{3}(x - 5) \Rightarrow 3y + 9 = -2x + 10,$$

which in general form is also $2x + 3y = 1$.

The correct answer is (B).

This problem can also be attacked by the trial-and-error approach. Plug in both points to each equation. Only one choice, (B), will satisfy both equations.

8. Let $f(x)$ be given by the left graph below and a transformation of f be given on the right graph below. What equation describes the graph of the transformation?

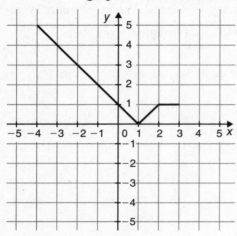
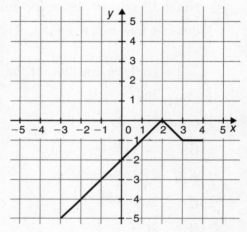

(A) $f(-x) + 1$
(B) $-f(x) - 1$
(C) $-f(x) + 1$
(D) $-f(x + 1)$
(E) $-f(x - 1)$

SOLUTION

Transformations are routine problems, but this problem involves a combination of transformations, making it nonroutine. There are two transformations here. The graph of $f(x)$ has been translated to the right one unit (so x must be replaced by $x - 1$) and it also has been rotated about the x-axis (so f becomes $-f$). (These transformations can be done in either order.) The correct answer is (E).

9. For the function $f(x) = 3^{x+2} + 2^{x+1}$, which of the following must be true?

I. The domain is $(-\infty, \infty)$.

II. The range is $(-\infty, \infty)$.

III. The y-intercept is 5.

(A) I only
(B) II only
(C) III only
(D) I and II only.
(E) I, II, and III

SOLUTION

This is a nonroutine problem that tests your knowledge of the meanings of domain, range, and y-intercept based on exponential functions. Statement I is true because you can raise numbers such as 3 and 2 to any power, so x can be any number. Statement II is not true because raising positive expressions to powers creates positive values, and two positives added cannot be 0 or less than 0, so this function would not be true for $f(x) \leq 0$. Statement III is not true because the y-intercept is the value of the function when $x = 0$, so the y-intercept would have to be $y = 3^2 + 2^1 = 11$. The correct answer is (A), I only.

Equations and Inequalities

10. Solve $|2x| + x < 3$.

(A) $(-1, 3)$
(B) $(-\infty, -3) \cup (1, \infty)$
(C) $(1, \infty)$
(D) $(-\infty, -3)$
(E) $(-3, 1)$

SOLUTION

This is a routine problem for solving absolute value inequalities.

$2x + x < 3$ $-(2x) + x < 3$

$3x < 3$ $-x < 3$

$x < 1$ $x > -3$

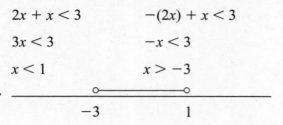

So $-3 < x < 1$, or $(-3, 1)$, and the correct answer is (E). You can quickly check that this is the correct interval by plugging $x = 0$ into the original inequality: $0 + 0 < 3$ is correct, and then substituting numbers not in the interval, such as $x = -4$ (which gives $4 < 3$, which is false) and $x = 2$ (which gives $4 + 2 > 3$, which is false).

Note that this problem can also be attacked by the trial-and-error approach by choosing a number in each interval choice and determining whether the inequality is satisfied inside the interval and not satisfied outside the interval. For example, choosing $x = 0$ in choice (A) yields a true statement, but then when checking outside of the interval, $x = -2$ is also true, which eliminates (A). This method is time-consuming, but it works. You must look at all possibilities when using trial-and-error in inequality problems.

11. Suppose the quadratic equation $x^2 + bx + 16 = 0$ has two solutions. If b is an integer, what is the smallest positive integer value for b and the largest negative integer value for b that satisfies this equation?

(A) smallest 9, largest -9

(B) smallest 8, largest -8

(C) smallest 5, largest -3

(D) smallest 4, largest -4

(E) smallest 17, largest 0

SOLUTION

This is a nonroutine problem. A routine problem would give you the equation and ask you how many solutions there were, forcing you to use the discriminant. But this problem asks you to find the value of b.

For the equation to have two real solutions, the discriminant must be positive.

Since $a = 1$ and $c = 16$,

$b^2 - 4ac = b^2 - 4(1)(16) > 0$

$b^2 - 64 > 0$

$(b + 8)(b - 8) > 0$

So both factors have to be positive or both factors have to be negative, which leads to

$b > 8$ or $b < -8$.

The smallest positive integer making the expression positive is 9, and the largest negative integer making the expression positive is –9.

The correct answer is (A).

12. Solve the system of equations: $\begin{cases} 5x - 2y + z = 14 \\ 6x + 4z = 10 \\ 8y + 5z = 6 \end{cases}$

(A) $x = 2, y = 0, z = -3$

(B) $x = -4, y = 2, z = 2$

(C) $x = 3, y = -\dfrac{1}{2}, z = -2$

(D) $x = 1, y = -7, z = 1$

(E) solution not given

SOLUTION

For the CLEP College Algebra exam, the preferred method for this problem would be to use trial-and-error and substitute the values of x, y, and z into each equation. All three equations must be true. The only answer choice that works is (C). You don't have to check (D) and (E).

$$\begin{cases} 5(3) - 2\left(\dfrac{-1}{2}\right) - 2 = 15 + 1 - 2 = 14 \\ 6(3) + 4(-2) = 18 - 8 = 10 \\ 8\left(\dfrac{-1}{2}\right) - 5(-2) = -4 + 10 = 6 \end{cases}$$

The technique for solving systems of three equations with three variables is not part of the CLEP College Algebra exam, but this nonroutine problem can be solved by substitution, similar to the technique for two equations with two unknowns. You can get x in terms of z from the second equation, and $2y$ in terms of z from the third equation:

$$x = \dfrac{5-2z}{3} \text{ and } 2y = \dfrac{6-5z}{4}.$$

Substitute these values into the first equation to get

$$5\left(\dfrac{5-2z}{3}\right) - \left(\dfrac{6+5z}{4}\right) + z = 14,$$ which has only one unknown, so you can find z. Multiply through by the LCD, 12, to get

$20(5 - 2z) - 3(6 + 5z) + 12z = 168$

$100 - 40z - 18 - 15z + 12z = 168$

$-43z = 86$

$z = -2$

Then substitute $z = -2$ into the second and third equations to get

$6x - 8 = 10$

$6x = 18$

$x = 3$

$8y + 10 = 6$

$8y = -4$

$y = -\dfrac{1}{2}$

The correct answer is (C).

Number Systems and Operations

13. Which of the following is *not* a rational number?

   I. $\left(\dfrac{\sqrt{3}}{\sqrt{2}}\right)^2$

   II. $-4.257257\ldots$

   III. $1.10100100001\ldots$

   (A) I only

   (B) II only

   (C) III only

   (D) I and III only

   (E) All are rational.

SOLUTION

This is a routine problem since you are responsible for classifying numbers. Realize first that you are asked about numbers that are *not* rational. Choice I, $\left(\dfrac{\sqrt{3}}{\sqrt{2}}\right)^2 = \dfrac{3}{2}$, is rational. Choice II is a repeating decimal, which is always rational. Choice III appears to be repeating as well, but looking at it carefully, although there is a pattern, you can see that there is no repetition, so III is irrational. The correct answer is (C).

14. Which of the following expressions equals 1?

(A) i^2

(B) i^3

(C) i^6

(D) i^8

(E) i^{25}

SOLUTION

This is a nonroutine problem. By combining the knowledge that $i = \sqrt{-1}$ and $i^2 = -1$ and the knowledge of raising a power to a power, you can evaluate all powers of i. So the values of (A) through (E) are the following:

(A) $i^2 = -1$; (B) $i^3 = i^2(i) = -i$; (C) $i^6 = i^3 (i^3) = (-i)(-i) = i^2 = -1$;
(D) $i^8 = (i^2)^4 = (-1)^4 = 1$; (E) $i^{25} = (i^4)^6 \, i = (1)^6 \, i$. The correct answer is (D).

15. Find $\sum_{n=1}^{5}(4^n - 1)$.

(A) 52

(B) 1023

(C) 1024

(D) 1359

(E) 1364

SOLUTION

This problem, if done by formula, could be considered a nonroutine one. Using formulas would require you to split this into two series, geometric and arithmetic, and then to use the formulas for the sums of the first five terms of each:

$$\sum_{n=1}^{5} 4^n - \sum_{n=1}^{5} 1 = \frac{4(1-4^5)}{1-4} - 5 = 1359.$$ The correct answer is (D).

But since there are only five terms to add, it is probably just as easy to generate the terms and add them using the scientific calculator that will be available to you: $(4^1 - 1) + (4^2 - 1) + (4^3 - 1) + (4^4 - 1) + (4^5 - 1)$
$= 3 + 15 + 63 + 255 + 1023 = 1359.$

The correct answer is (D). This approach makes the problem a routine one because it simply tests your knowledge of sigma notation.

16. If the 4th terms of $(2x + y)^6$ and $(2x - y)^6$ were added, their sum would be:

(A) 0
(B) $160x^3y^3$
(C) $320x^3y^3$
(D) $160x^4y^2$
(E) $320x^4y^2$

SOLUTION

This is a nonroutine problem in that you need to use the binomial theorem twice. However, if you realize that the 4th term of $(2x + y)^6$ is positive and the 4th term of $(2x - y)^6$ is negative with the same coefficients and powers of x and y, the problem is easy. The answer is zero, and (A) is correct.

If you do this problem in a straightforward way, you can use either the binomial theorem or Pascal's triangle to get the coefficients. The x and y terms are found the same way for both methods, with $n = 6$ and $r = 4$. The exponent of the y term is $r - 1$, so it is 3, and the exponent of the x term is $n -$ (the y exponent), or $6 - 3 = 3$. So the term will contain $(2x)^3y^3 = 8x^3y^3$ as factors.

Now, to find the coefficient.

Using the binomial theorem and the combination formula: $_nC_{r-1} = \dfrac{n!}{(r-1)!(n-r+1)!}$, the coefficient of the 4th term is

$_6C_3 \dfrac{6!}{(3)!(3)!} = \dfrac{6(5)(4)}{3(2)(1)} = 20.$

Using Pascal's triangle, since it is the 4th term, you need to write only the first four terms of any row:

Row 0				1		
Row 1				1	1	
Row 2			1	2	1	
Row 3		1	3	3	1	
Row 4		1	4	6	4	
Row 5	1	5	10	10		
Row 6	1	6	15	20		

The result is, again, 20.

So the 4th term of $(2x + y)^6$ is $20(8x^3)y^3$, and the 4th term of $(2x - y)^6$ is $-20(8x^3)y^3$. The sum of these terms is zero and the correct answer is (A).

FORMULAS YOU MUST KNOW

It should be obvious that the nonroutine problems are more difficult than the routine problems. So you need to maximize your score as much as possible in the approximately 30 routine problems that you will face on the exam. If you can do a majority of them, you have more leeway in the nonroutine problems.

The best way to maximize your score is to know all of the important formulas that you are bound to see. Many routine problems require knowledge of these formulas. Here are the important formulas from Chapters 3 through 6.

Basic Algebra

Description	Formula
Additive inverse to a	$-a$
Multiplicative inverse to a ($a \neq 0$)	$\dfrac{1}{a}$
Raising expression to a negative exponent	$b^{-n} = \dfrac{1}{b^n}$
Multiplying expressions with powers	$x^m \cdot x^n = x^{m+n}$
Raising expressions to powers	$(x^m)^n = x^{mn}$
Dividing expressions with powers	$\dfrac{x^m}{x^n} = x^{m-n}$
Fractional exponents	$x^{1/2} = \sqrt{x}$, $x^{1/3} = \sqrt[3]{x}$, $x^{m/n} = \sqrt[n]{x^m} = \left(\sqrt[n]{x}\right)^m$
Multiplying radicals	$\sqrt{a} \cdot \sqrt{b} = \sqrt{ab}$
Using FOIL to multiply binomials	$(ax + b)(cx + d) = acx^2 + adx + bcx + bd$ FOIL means to multiply the terms in the order "First, Outer, Inner, Last".
Factoring: difference of 2 squares	$x^2 - y^2 = (x + y)(x - y)$
Factoring: perfect squares	$x^2 + 2xy + y^2 = (x + y)^2$ $x^2 - 2xy + y^2 = (x - y)^2$
Factoring: sum and difference of 2 cubes	$x^3 + y^3 = (x + y)(x^2 - xy + y^2)$ $x^3 - y^3 = (x - y)(x^2 + xy + y^2)$
Multiplying rational expressions	$\dfrac{a}{b} \times \dfrac{c}{d} = \dfrac{ac}{bd}$
Dividing rational expressions	$\dfrac{a}{b} \div \dfrac{c}{d} = \dfrac{a}{b} \times \dfrac{d}{c} = \dfrac{ad}{bc}$
Finding conjugates	conjugate of $\sqrt{a} + \sqrt{b} = \sqrt{a} - \sqrt{b}$ conjugate of $\sqrt{a} - \sqrt{b} = \sqrt{a} + \sqrt{b}$

Functions and Their Properties

Description	Formula
Slope of line	Given two points (x_1, y_1) and (x_2, y_2), $m = \dfrac{\text{rise}}{\text{run}} = \dfrac{y_2 - y_1}{x_2 - x_1}$.
Equation of line: slope-intercept form	$y = mx + b$, where m is the slope, b is the y-intercept
Equation of line: point-slope form	Given two points (x_1, y_1) and (x_2, y_2), $(y - y_1) = m(x - x_1)$, where $m = \dfrac{y_2 - y_1}{x_2 - x_1}$, as above.
Equation of line: intercept form	Given x-intercept a and y-intercept b, $\dfrac{x}{a} + \dfrac{y}{b} = 1$
Slopes of parallel lines (equal)	$m_1 = m_2$
Slopes of perpendicular lines (negative reciprocals)	$m_2 = \dfrac{-1}{m_1}$ or $m_1 \cdot m_2 = -1$

Equations and Inequalities

Description	Formula
Absolute value	$\lvert a \rvert = \begin{cases} a, & a \geq 0 \\ -a, & a < 0 \end{cases}$
Quadratic formula	For $ax^2 + bx + c = 0$, $x = \dfrac{-b \pm \sqrt{b^2 - 4ac}}{2a}$.
Discriminant used to determine the number of real solutions	$b^2 - 4ac \begin{cases} > 0, \ 2 \text{ real solutions} \\ = 0, \ 1 \text{ real solution} \\ > 0, \ 2 \text{ real solutions} \\ \quad \text{(or 2 imaginary)} \end{cases}$
Logarithm	$\log_b y = x$ is equivalent to $b^x = y$
Common logarithm	$\log y = x$ is equivalent to $10^x = y$
Natural logarithm	$\ln y = x$ is equivalent to $e^x = y$ ($e = 2.71828...$)
Logarithm addition rule (with any base)	$\log a + \log b = \log(a \cdot b)$
Logarithm subtraction rule (with any base)	$\log a - \log b = \log\left(\dfrac{a}{b}\right)$
Logarithm power rule (with any base)	$\log a^b = b \log a$

Number Systems and Operations

Description	Formula
Imaginary unit	$i = \sqrt{-1}$ and $i^2 = -1$
Complex conjugate	Complex conjugate of $a + bi$ is $a - bi$. Complex conjugate of $a - bi$ is $a + bi$.
Arithmetic sequence	$a_n = a_1 + (n-1)d$, where a_1 is the first term, d is the common difference, and n is the number of terms
Sum of arithmetic sequences (arithmetic series)	$S_n = \dfrac{n}{2}[2a_1 + (n-1)d]$ or $S_n = \dfrac{n}{2}(a_1 + a_n)$
Geometric sequence	$a_n = a_1 r^{n-1}$, where a_1 is the first term, r is the common ratio and n is the number of terms
Sum of geometric sequences (geometric series)	$S_n = \dfrac{a_1(1-r^n)}{1-r}$, where a_1 is the first term, r is the common ratio, and n is the number of terms
Sum of infinite geometric series	$S_n = \dfrac{a_1}{1-r}$, where a_1 is the first term, r is the common ratio, and $|r| < 1$. If $|r| \geq 1$, the series diverges.
Factorial	$n! = n(n-1)(n-2)\ldots 3(2)(1)$
Combination of n choose r	$_nC_r = \dfrac{n!}{r!(n-r)!}$
Binomial theorem	$(x+y)^n = {_nC_0}x^n + {_nC_1}x^{n-1}y + {_nC_2}x^{n-2}y^2 + {_nC_3}x^{n-3}y^3 + \ldots + {_nC_{n-2}}x^2y^{n-2} + {_nC_{n-1}}xy^{n-1} + {_nC_n}y^n$
Determinant of 2-by-2 matrix	$\begin{vmatrix} a & b \\ c & d \end{vmatrix} = ad - bc$

WHEN YOU SEE THE WORDS...

As difficult as mathematics is for many students, its one advantage is that there are specific tasks you know you will have to perform. You know you will have to simplify expressions, factor, solve equations, and graph functions. And that makes studying for the CLEP College Algebra exam easier.

It is difficult to predict every type of problem that you will have on the CLEP College Algebra exam, especially nonroutine problems, but some types of problems almost always show up on the exam with specific wording.

This section looks at the problem types that you are most likely to see, how they are worded, and how you should attack them.

Algebraic Operations

General problem wording	Step(s) to follow
Find the additive inverse of (*expression*).	Solve the equation for y: $y + (expression) = 0$. It is the negative of the expression: $-(expression)$.
Find the multiplicative inverse of (*expression*).	Solve the equation for y: $y(expression) = 1$. It is the reciprocal of the *expression*: $\dfrac{1}{(expression)}$
Raise (*expression*) to a negative power.	$(expression)^{-n} = \dfrac{1}{(expression)^n}$.
Raise (*expression*) to a fractional power.	$(expression)^{\frac{1}{n}} = \sqrt[n]{expression}$.
Fully factor (*expression*).	Write (*expression*) as a multiplication of factors, none of which can be further factored.
Add or subtract algebraic fractions.	Factor all denominators and find a common denominator by multiplying denominators. Multiply each term by that common denominator and add the terms. All terms should then have the same denominator, so you just have to add or subtract the numerators.

(continued)

(continued)

General problem wording	Step(s) to follow
Simplify a complex fraction.	Find the least common denominator for all fractions and multiply every term by that LCD.
Rationalize $\sqrt{\dfrac{a}{b}}$.	$\sqrt{\dfrac{a}{b}} = \dfrac{\sqrt{a}}{\sqrt{b}}$. Then multiply numerator and denominator by $\sqrt{b}$: $\sqrt{\dfrac{a}{b}} = \dfrac{\sqrt{a}}{\sqrt{b}}\left(\dfrac{\sqrt{b}}{\sqrt{b}}\right) = \dfrac{\sqrt{a}}{b}$.
Solve a linear equation.	Simplify the equation to get all unknowns (x's) on one side of the equation and all other terms on the other side. Divide both sides of the equation by the coefficient of x, and simplify. There can be only one solution.
Solve an equation with a square root in it.	Move the square root term to one side of the equation and everything else to the other side. Square both sides and solve. Make sure you check the solution in the original equation because squaring may introduce extraneous answers.

Functions and Their Properties

General problem wording	Step(s) to follow
Determine whether a graph is a function.	Apply the vertical line test. Draw vertical lines at various locations and check to be sure none intersect the graph in more than one location.
Determine whether a given graph matches that of an algebraic function.	By trial and error, plug values for x into the algebraic function, find the value of y, and determine whether those points are on the graph.
Given a graph of $f(x)$, find $f(f(a))$.	Find the value of y at $x = a$. Then find the value of y at $x =$ the answer you just found.
Given an algebraic function $y = f(x)$, find its domain.	The domain will be (∞, ∞) (all numbers) with the following exceptions: (a) any value of x that makes the denominator 0, (b) any value of x that makes an even root of a negative number; (c) any value of x that makes a term into the log of zero or the log of a negative number.
Given an algebraic function $y = f(x)$, find its range.	The range will be (∞, ∞) (all numbers) with the following exceptions: (a) any function in the form of x^n where n is an even number will have a range $[0, \infty)$ (all non-negative real numbers); (b) any function in the form of $\sqrt[n]{x}$ where n is an even number will have a range of $[0, \infty)$ (all non-negative real numbers); (c) any function in the form of a^x will have a range of $(0, \infty)$ (all positive real numbers).
Graph a line in the form $y = mx + b$.	Choose two values for x and find the corresponding values of y. Draw a line between both points.

(continued)

(continued)

General problem wording	Step(s) to follow
Given points (x_1, y_1) and (x_2, y_2), find the slope of the line through the points.	$m = \dfrac{\text{rise}}{\text{run}} = \dfrac{y_2 - y_1}{x_2 - x_1}$.
Find the equation of the line with slope m and y-intercept b.	Use the slope-intercept form: $y = mx + b$.
Find the equation of the line passing through the point (x_1, y_1) with slope m.	Use the point-slope form: $y - y_1 = m(x - x_1)$.
Find the equation of the line passing through two points.	Find slope m and then use the point-slope form $y - y_1 = m(x - x_1)$, using either point as (x_1, y_1)
Find the equation of the line with x-intercept a and y-intercept b.	Either find the line passing through the points $(a, 0)$ and $(0, b)$ as above or use the intercept form: $\dfrac{x}{a} + \dfrac{y}{b} = 1$.
Find the equation of the line parallel to line L passing through point $P(x_1, y_1)$.	Put line L in the form $y = mx + b$, where m is the slope, and then use the point-slope form $y - y_1 = m(x - x_1)$ with the given point.
Find the equation of the line perpendicular to line L passing through point $P(x_1, y_1)$.	Put line L in the form $y = mx + b$, where m is the slope, and then the slope of the perpendicular line is $\dfrac{-1}{m}$. Use the point-slope form with the given point and new slope $\dfrac{-1}{m}$, so $y - y_1 = -\dfrac{1}{m}(x - x_1)$.
Graph an inequality in the form of $y > mx + b$, $y \geq mx + b$, $y < mx + b$, or $y \leq mx + b$.	Graph the equation $y = mx + b$. If the inequality sign is $\geq$ or $\leq$, the line will be solid; otherwise it will be dashed. If the inequality sign is $>$ or $\geq$, shade the area above the line. If the inequality sign is $<$ or $\leq$, shade the area below the line.

(continued)

General problem wording	Step(s) to follow
Determine whether a function given in graph form is even, odd, or neither.	If the graph is symmetric with respect to the y-axis (quadrants I and II, III and IV are mirror images), the function is even. If the graph is symmetric with respect to the origin (quadrants I and III, II and IV are mirror images), the function is odd. Otherwise, it is neither.
Given the graph of $f(x)$, describe the basic transformations of the graph.	$f(x) + a$: shift graph a units up; $f(x) - a$: shift graph a units down; $f(x - a)$: shift graph a units right; $f(x + a)$: shift graph a units left; $-f(x)$: reflect graph across x-axis.
Given a graph, determine the number of roots.	Determine how many times the graph touches or crosses the x-axis.
Given a polynomial of degree n, find the maximum and minimum number of real roots.	The maximum number of real roots is n. If n is even, the minimum number of real roots is zero. If n is odd, the minimum number of real roots is one.
Given a function $f(x)$, determine the roots.	Set $f(x) = 0$ and factor $f(x)$. Set each factor equal to zero and solve, if possible. Also, a root occurs in the interval (a, b) if $f(a)$ and $f(b)$ have opposite signs.
Find the inverse f^{-1} of a given function $y = f(x)$	Interchange x and y and solve for y.
Given the graph of a function $f(x)$, determine whether its inverse $f^{-1}(x)$ is a function.	Apply the horizontal line test on $f(x)$. Draw a horizontal line at various locations. If it doesn't interest the graph at more than one location, the inverse $f^{-1}(x)$ is a function.

Equations and Inequalities

General problem wording	Step(s) to follow
Solve an inequality in the form of $a < f(x) < b$.	Split the problem into two inequalities: $a < f(x)$ and $f(x) < b$. The solution will be an interval common to both.
Solve an absolute value equation in the form of $\|f(x)\| = a$.	Split the problem into two equations: $f(x) = a$ and $-f(x) = a$. Solve each and check by plugging the answers into the original equation since one or both solutions could be extraneous.
Solve an absolute value equation in the form of $\|f(x)\| < a$ or $\|f(x)\| > a$.	Change the absolute value to two inequalities: $f(x) < a$, and $-f(x) < a$, or $f(x) > a$, and $-f(x) > a$, depending on the original inequality. Create a number line with the two solutions creating intervals. Check numbers in each interval to see whether it makes the absolute value inequality true.
Solve a quadratic equation in the form of $ax^2 + bx + c = 0$.	Either factor the quadratic and set each factor equal to zero and solve; or use the quadratic formula $x = \dfrac{-b \pm \sqrt{b^2 - 4ac}}{2a}$.
Determine the number of real roots for a quadratic equation $ax^2 + bx + c = 0$	Examine the discriminant: $b^2 - 4ac$. $$b^2 - 4ac \begin{cases} > 0, 2 \text{ real solutions} \\ = 0, 1 \text{ real solution} \\ = 0, 0 \text{ real solutions} \\ \quad \text{(or 2 imaginary solutions)} \end{cases}$$
Given a quadratic equation in the form of $ax^2 + bx + c = 0$, where a and b are given, find values of c for which the quadratic has 2, 1, or no solutions.	Use the discriminant $b^2 - 4ac$ and solve for c: $b^2 - 4ac > 0$ for 2 solutions $b^2 - 4ac = 0$ for 1 solution $b^2 - 4ac < 0$ for no solutions This method works equally well when a and c are given and b is to be determined.

(continued)

(continued)

General problem wording	Step(s) to follow
Solve a quadratic inequality in the form of $ax^2 + bx + c > 0$ or $ax^2 + bx + c < 0$.	Treat the quadratic as an equation and solve using techniques described above. Then create a number line with the two solutions creating intervals and check numbers in each interval to see which make the quadratic inequality true.
Solve a system of equations in the form of $\begin{cases} ax + by = c \\ dx + ey = f \end{cases}$.	Substitution method: If it is easy to do so, solve for a variable in one equation, then plug it into the second equation and solve. Knowing one variable allows you to use the other equation to find the second variable. Elimination method: Multiply both equations by numbers that will allow you to add the two equations with one variable canceling out. Solve the resultant equation. Knowing one variable allows you to use either original equation to find the other variable.
Solve a system of inequalities graphically.	Graph each inequality using techniques for graphing a line, shading the solution regions. The solution will be the region that has a common shading.
Determine whether a function is a growth or decay curve.	When the equation is in the form of $y = a^x$, the graph is a growth curve if $a > 1$ and a decay curve if $0 < a < 1$.
Solve an equation in the form of $a^x = b$.	Express a and b as powers of the same base, if possible. Then set the exponents equal to each other and solve for x.
Find the value of $\log_b y$.	Set $\log_b y = x$. Then write the equivalent equation $b^x = y$ and solve as above.
Find the value of $\log y$.	Set $\log y = x$ and solve the equation $10^x = y$ as above.

(continued)

(continued)

General problem wording	Step(s) to follow
Find the value of ln y.	Set ln $y = x$ and solve the equation $e^x = y$, where $e = 2.71828...$. Most natural log answers are left in terms of e.
Simplify logarithm expressions (any base).	$\log a + \log b = \log(a \cdot b)$ $\log a - \log b = \log\left(\dfrac{a}{b}\right)$ $\log a^b = b \log a$
Solve logarithmic equations in the form of $\log_b f(x) = n$.	Rewrite the equation exponentially: $b^n = f(x)$ and solve.

Number Systems and Operations

General problem wording	Step(s) to follow
Which of the following numbers are rational...	Rational numbers include integers, fractions, terminating decimals, and repeating decimals.
Use imaginary number notation for $\sqrt{-n}$	$\sqrt{-n} = \sqrt{-1}\sqrt{n} = i\sqrt{n}$
Multiply two complex numbers $(a + bi)(c + di)$.	Use FOIL techniques, remembering that $i^2 = -1$.
Simplify an expression in the form of $\dfrac{a + bi}{ci}$.	Multiply the fraction by $\dfrac{i}{i}$ and simplify both numerator and denominator.
Simplify an expression in the form of $\dfrac{a + bi}{c + di}$.	Multiply the fraction by the complex conjugate $\dfrac{c - di}{c - di}$ and simplify both numerator and denominator.
Given a formula for a_n, find the nth term of a sequence.	Plug n into the formula for a_n.
Given a formula for a_n, find the partial sum S_n of a series.	If n is small, generate n terms and add them. Or use the formula for S_n.
Find $\sum\limits_{n=1}^{k} a(n)$.	This is a series: $a(1) + a(2) + ... + a(k)$. Generate each term and add them. Or use the formula for S_n.

(continued)

(continued)

General problem wording	Step(s) to follow
Determine whether a sequence is arithmetic, geometric, or neither.	Subtract the first term from the second, the second term from the third and so on. If you get a common difference d, it is arithmetic. Divide the second term by the first, the third term by the second, and so on. If you get a common ratio r, it is geometric.
Given the first term a_1, the common difference d, and a value for n, find the nth term of an arithmetic sequence.	Use the formula $a_n = a_1 + (n-1)d$.
Find the sum of the first n terms of an arithmetic series (the sum of a sequence).	If you know the first term a_1, the common difference d, and a value for n, use $$S_n = \frac{n}{2}[2a_1 + (n-1)d].$$ If you know the first term a_1, the nth term a_n, and n, use $$S_n = \frac{n}{2}(a_1 + a_n).$$
Given the first term a_1, the common ratio r, and a value for n, find the nth term of a geometric sequence.	Use the formula $a_n = a_1 r^{n-1}$.
Find the sum of the first n terms of a geometric series (the sum of a geometric sequence).	Generate the terms and add them, or use the formula $S_n = \dfrac{a_1(1-r^n)}{1-r}$
Find the sum of an infinite geometric series.	If $|r| < 1$, $S_n = \dfrac{a_1}{1-r}$. Otherwise, the sum is infinite.
Calculate $n!$.	$n! = n(n-1)(n-2)...(3)(2)(1)$
Calculate $_nC_r$.	$_nC_r = \dfrac{n!}{r!(n-r)!}$

(continued)

(continued)

General problem wording	Step(s) to follow
Expand $(x+y)^n$ and $(x-y)^n$ by using the binomial theorem.	$(x+y)^n = {}_nC_0 x^n + {}_nC_1 x^{n-1}y + {}_nC_2 x^{n-2}y^2 + \ldots + {}_nC_{n-2}x^2 y^{n-2} + {}_nC_{n-1}xy^{n-1} + {}_nC_n y^n$. The $(x-y)^n$ expansion is the same except all the even terms are negative.
Find the rth term of $(x+y)^n$ and $(x-y)^n$.	The coefficient of the term is ${}_nC_{r-1}$. The x-variable: x^{n-r+1}. The y-variable: y^{r-1}. For $(x-y)^n$, if r is even, the term is negative.

COMMON MISTAKES TO AVOID

When you multiply fractions, you multiply numerators and denominators. But when you add fractions, you do not add numerators and denominators. Algebra rules can be confusing and students may use rules that make sense to them, but are completely incorrect. You should remember that the word "term" implies addition or subtraction, and the word "factor" implies multiplication. The lists below point out examples of the most common mistakes students make and how to avoid them.

Errors Involving Parentheses

This is WRONG!	This is RIGHT!	Comment
$4 - (x - y) = 4 - x - y$	$4 - (x - y)$ $= 4 - x - (-y)$ $= 4 - x + y$	Distribute negative sign to all terms within parentheses.
$(x + 5)^2 = x^2 + 25$	$(x + 5)^2 = x^2 + 10x + 25$	This is a FOIL problem.
$(4x - 8)^2 = 4(x - 2)^2$	$(4x - 8)^2$ $= [4(x - 2)]^2$ $= 16(x - 2)^2$	When factoring, the exponent applies to both factors.

Errors Involving Fractions

This is WRONG!	This is RIGHT!	Comment
$\dfrac{x+2}{2} = x+1$	Leave $\dfrac{x+2}{2}$ alone!	Cannot cancel terms in fractions unless they all have a common factor.
$\dfrac{2}{x+2} = \dfrac{1}{x}+1$	Cannot simplify $\dfrac{2}{x+2}$.	Cannot split a fraction with a binomial denominator.
$\dfrac{1}{x}+\dfrac{1}{4} = \dfrac{2}{x+4}$	$\dfrac{1}{x}\left(\dfrac{4}{4}\right)+\dfrac{1}{4}\left(\dfrac{x}{x}\right) =$ $\dfrac{4}{4x}+\dfrac{x}{4x} = \dfrac{x+4}{4x}$	To add fractions, you need a common denominator.
$\dfrac{1}{5x} = \dfrac{1}{5}x$	$\dfrac{1}{5x} = \dfrac{1}{5}\left(\dfrac{1}{x}\right)$	The x is in the denominator.
$(1/5)x = \dfrac{1}{5x}$	$(1/5)x = \dfrac{x}{5}$	The x is in the numerator. Avoid fractions with slanted lines. They are confusing.

Errors Involving Radicals

This is WRONG!	This is RIGHT!	Comment
$\sqrt{49} = \pm 7$	$\sqrt{49} = 7$	Square roots are always positive. The equation $x^2 = 49$ has solutions of $x = \pm 7$, however.
$\sqrt{10x} = 10\sqrt{x}$	$\sqrt{10x} = \sqrt{10}\sqrt{x}$	Apply radical to every term inside the radical sign.
$\sqrt{\dfrac{5x}{4}} = \dfrac{\sqrt{5x}}{4}$	$\sqrt{\dfrac{5x}{4}} = \dfrac{\sqrt{5x}}{\sqrt{4}} = \dfrac{\sqrt{5x}}{2}$	To take square root of fractions, take square root of both numerator and denominator.
$\sqrt{x^2 + 36} = x + 6$	Cannot simplify $\sqrt{x^2 + 36}$	Cannot apply the square root separately to each term.
$\sqrt{-4} \cdot \sqrt{-25} = \sqrt{100} = 10$	$\sqrt{-4} \cdot \sqrt{-25}$ $= \sqrt{4}\sqrt{-1} \cdot \sqrt{25}\sqrt{-1}$ $= 2i(5i) = 10i^2 = -10$	When working with imaginary numbers, write each radical in terms of i.

Errors Involving Exponents

This is WRONG!	This is RIGHT!	Comment
$x^2 + x^4 = x^6$	Cannot simplify $x^2 + x^4$	When adding terms, variables and *exponents* must be the same.
$x^2(x^4) = x^8$	$x^2(x^4) = x^6$	When multiplying expressions with the same base, add the exponents.
$4^2(4^4) = 16^6$	$4^2(4^4) = 4^6$	When multiplying expressions with the same base, the base doesn't change.
$(x^3)^3 = x^{27}$	$(x^3)^3 = x^9$	When raising expressions to powers, multiply the exponents; the base doesn't change.
$(5x^4)^3 = 5x^{12}$	$(5x^4)^3 = 5^3 x^{12} = 125x^{12}$	When raising expressions to powers, every factor gets raised to that power.
$4^{-2} = -8$	$4^{-2} = \dfrac{1}{4^2} = \dfrac{1}{16}$	Negative exponents do not create negative numbers, they create fractions.
$(4x)^{-1} = \dfrac{4}{x}$	$(4x)^{-1} = \dfrac{1}{(4x)^1} = \dfrac{1}{4x}$	The parentheses indicate that the exponent applies to both the 4 and the *x*.
$\dfrac{1}{x^5 + x^3} = x^{-5} + x^{-3}$	Cannot simplify $\dfrac{1}{x^5 + x^3}$	Cannot move terms from denominator to numerator, only factors.

Miscellaneous Errors

This is WRONG!	This is RIGHT!	Comment
$f(g(2)) = f(2) \cdot g(2)$	The notation $f(g(2))$ does not indicate multiplication—it is a composite function.	To find $f(g(2))$, plug 2 into the g function, and plug that answer into the f function.
If $-4x > 8$ then $x > -2$	If $-4x > 8$ then $x < -2$	When solving inequalities, if you multiply or divide by a negative number, the direction of the inequality sign reverses.
Horizontal lines have no slope.	Horizontal lines have slope $= 0$.	Vertical lines have no slope.
The slope of the line passing through $(4, 5)$ and $(3, 2)$ is $\dfrac{4-3}{5-2} = \dfrac{1}{3}$	Slope $= \dfrac{5-2}{4-3} = 3$	Slope $= \dfrac{\text{rise}}{\text{run}} = \dfrac{\text{change in } y}{\text{change in } x}$.
The number of real roots of a polynomial is the degree of the polynomial.	The *maximum* number of real roots of a polynomial is the degree of the polynomial.	Real roots occur when the graph of the polynomial touches or crosses the x-axis. Other roots can be imaginary.
The inverse to $y = 2x$ is $y = \dfrac{1}{2x}$	The inverse to $y = 2x$ is $x = 2y$, or $y = \dfrac{x}{2}$	The reciprocal to $y = 2x$ is $y = \dfrac{1}{2x}$. To find inverses, interchange x and y.
$\log_2(-8) = -3$	$\log_2(-8)$ doesn't exist	$\log_2(-8) = x \Rightarrow 2^x = -8$. but 2 raised to any power must be positive. You cannot take an even-base log of a negative number.

REVIEW OF TEST-TAKING TIPS

In General

Become comfortable with the format of the exam. Stay calm and pace yourself. Simulating the test a few times with the practice tests will boost your confidence.

Work quickly and steadily. You have on average about 1.5 minutes to answer each problem. Be aware of your time. Spending 10 minutes on a problem and ultimately getting it correct might mean that you might not complete the exam. If you feel that you are spending too much time on a problem, simply guess (eliminating any choices you know are wrong) and move on. If you have time left, come back to the problem.

Beware of test vocabulary. Words such as *always, every, none, only,* and *never* indicate there should be no exceptions to the answers you choose. Words such as *generally, usually, sometimes, seldom, rarely,* and *often* indicate that there may be exceptions to your answers.

Multiple-Choice Problems

Be neat. Many students take the philosophy that since their work won't be seen, they can be sloppy. Sloppy work, though, leads to careless mistakes. Also, if you come back to the problem later in the exam, you want to look at work that is easy to understand.

Once you start a problem, make sure you answer it. Eliminate any incorrect choices and answer the question the best you can. If you are unsure, write down the problem number on your scrap paper and then go back if you have time.

Develop a set of simple codes that tell you what to do if you are not sure of how to solve a problem. For instance, on your scrap paper, you might want to circle the problem number if you feel that more time would help you; if you finish the exam early, you can return to that problem. Putting an X through a problem number might mean that you should not spend any more time on the problem.

Be sure you answer every question. Don't leave any question blank, even if you have to guess randomly. Try to eliminate some answer choices first. Only your correct answers count.

Get right what you are sure you know. If you see a problem and you absolutely know how to solve it, you must get it correct, even if you need to spend some extra time on it. Every correct problem is gold!

Read all the possible answers. Just because you think you have found the correct response, do not automatically assume that it is the best answer. Read through each choice to be sure that you are not making a mistake by jumping to conclusions.

Use the process of elimination. Even if you do not know how to solve a problem, go through each answer to a question and eliminate as many of the answer choices as possible. For instance, if you were asked to find the minimum value of $x^2 + 4$, it would be impossible for the answer to be a negative number as the smallest possible value for x^2 is zero.

Read the answers carefully. The equations $2x + 3y = 4$ and $2x - 3y = 4$ look quite similar. When you have answers that look similar, be sure you carefully choose the correct one.

Remember that every answer choice is independent. Just because you have answered four questions in a row with the answer (C) doesn't mean that the next answer *cannot* be (C). Just because you haven't had the answer (E) for a long time doesn't mean that it is time for an answer to be (E).

Answer the question that is asked. If you were asked to find the value of x^2 when $x = \log_3 81$, be sure to choose the correct answer of $x^2 = 16$ rather than the value of x, which is 4. *The answer to an equation is not necessarily the answer to a problem.* Taking a second to underline the actual question being asked will help to eliminate the possibility of errors such as this.

Be absolutely aware of the time. Wear a watch, and when the exam starts, jot down the time and, more important, the time when the exam ends, 90 minutes after the start time. If time is becoming an issue, start looking through the remaining problems, and answer the ones you know how to solve. With 5 minutes left, go back and make sure you have an answer to every problem, even those you haven't attempted.

PRACTICE TEST 1
CLEP College Algebra

Also available at the REA Study Center (*www.rea.com/studycenter*)

> This practice test is also offered online at the REA Study Center. Since all CLEP exams are administered on computer, we recommend that you take the online version of this test to simulate test-day conditions and to receive these added benefits:
>
> - **Timed testing conditions** – helps you gauge how much time you can spend on each question
> - **Automatic scoring** – find out how you did on the test, instantly
> - **On-screen detailed explanations of answers** – gives you the correct answer and explains why the other answer choices are wrong
> - **Diagnostic score reports** – pinpoint where you're strongest and where you need to focus your study

PRACTICE TEST 1

CLEP College Algebra

(Answer sheets appear in the back of the book.)

TIME: 90 Minutes
60 Questions

DIRECTIONS: An online scientific calculator will be available for the questions in this test.

Some questions will require you to select from among five choices. For these questions, select the BEST of the choices given.

Some questions will require you to type a numerical answer in the box provided.

Notes:

(1) Unless otherwise specified, the domain of any function f is assumed to be the set of all real numbers x for which $f(x)$ is a real number.

(2) i will be used to denote $\sqrt{-1}$.

(3) Figures that accompany questions are intended to provide information useful in answering the questions. All figures lie in a plane unless otherwise indicated. The figures are drawn as accurately as possible *except* when it is stated in a specific question that the figure is not drawn to scale. Straight lines and smooth curves may appear slightly jagged on the screen.

1. The number 4.563563... fits into which classifications?
    I. Complex
    II. Real
    III. Rational

    (A) I only
    (B) II only
    (C) III only
    (D) I and II only
    (E) I, II, and III

2. Solve the inequality $-9 \leq -4x - 3 < 1$.

   (A) $(-1, \frac{3}{2}]$
   (B) $[-1, \frac{3}{2})$
   (C) $(-3, -1)$
   (D) $(-3, -1]$
   (E) $[-3, -1)$

3. If $f(x) = \frac{1}{x-4}$ and $g(x) = x^2$, what values of x are NOT in the domain of $f(g(x))$?

   (A) 2 only
   (B) 4 only
   (C) $-2, 2$ only
   (D) $-2, 2, 4$
   (E) All values of x are in the domain of f.

4. The equation that has the greatest number of real roots is

   (A) $4x - 5 = 0$
   (B) $(x + 3)^2 = 0$
   (C) $x^2 + 1 = 0$
   (D) $x^3 - 1 = 0$
   (E) $x^2 + 3x + 1 = 0$

5. Solve the equation $|2x - 1| - 3x = 9$ for x.

   (A) $x = \frac{-8}{5}$ only
   (B) $x = -10$ only
   (C) $x = \frac{-8}{5}, x = -10$
   (D) $x = -2$ only
   (E) $x = -2, x = -10$

6. What is the value of $\sum_{n=0}^{5}(2+x^2-2^x)$?

   ☐

7. Find the value of $\log_2 8 + \log_8 2$.

   (A) 0
   (B) 1
   (C) $\dfrac{10}{3}$
   (D) $\dfrac{13}{4}$
   (E) $\dfrac{17}{4}$

8. The function $y = x^3 - 5x^2 + 6x + 1$ must have a root between which of the following x-values?

   (A) −2 and −1
   (B) −1 and 0
   (C) 0 and 1
   (D) 1 and 2
   (E) 2 and 3

9. The system of equations $\begin{cases} y=(x-1)^2 \\ y=(x+1)^2 \end{cases}$ has how many points of intersection?

   (A) 0
   (B) 1
   (C) 2
   (D) 3
   (E) 4

10. If both x and y are not zero, what is the multiplicative inverse to the expression $\dfrac{x^{-2}}{y^{-1}}$?

   (A) $\dfrac{x^2}{y}$

   (B) $\dfrac{-x^2}{y}$

   (C) $\dfrac{y}{x^2}$

   (D) $\dfrac{y}{-x^2}$

   (E) $x^2 y$

11. Find the equation of the line passing through the points $(2, -1)$ and $(4, 2)$.

   (A) $2x - 3y = 2$
   (B) $2x + 3y = 14$
   (C) $3x = 2y$
   (D) $3x - 2y = 8$
   (E) $3x + 2y = 16$

12. If $f(x) = x^2 - 4x + 10$ and $g(x) = |2 - x|$, what is the value of $f(g(5))$?

   ☐

13. What is the domain of $y = \dfrac{\sqrt{x+2}}{x+1}$?

   (A) $[-2, -1)$
   (B) $(-2, \infty)$
   (C) $[-2, \infty)$
   (D) $(-2, -1) \cup (-1, \infty)$
   (E) $[-2, -1) \cup (-1, \infty)$

14. Which of the following has the largest value?

(A) 4!

(B) $\dfrac{6!}{4!}$

(C) $\dfrac{7!}{5!\,2!}$

(D) $\dfrac{8!}{4!\,4!\,2!}$

(E) $\dfrac{20!}{19!}$

15. Which of the following graphs is NOT a function?

(A)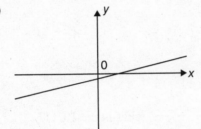

(B)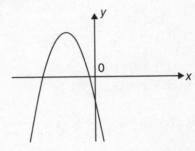

(C)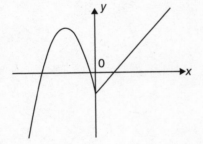

(D)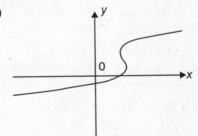

(E)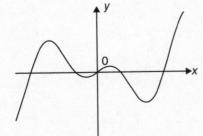

16. $\dfrac{3+i}{1-i}$ is equivalent to

   (A) $1 + 2i$
   (B) $1 + i$
   (C) $3 - i$
   (D) 2
   (E) 3

17. Solve for x: $4^{x-1} = \log_8 64$

   (A) $x = \dfrac{1}{2}$
   (B) $x = \dfrac{3}{2}$
   (C) $x = \dfrac{5}{2}$
   (D) $x = 1$
   (E) $x = 3$

18. To simplify $\dfrac{x}{2x+6} - \dfrac{3}{3x-9} + \dfrac{5}{4x^2+12x}$, what is the LCD?

   (A) $24(x + 3)(x - 3)$
   (B) $12(x + 3)^2(x - 3)$
   (C) $12x(x + 3)(x - 3)$
   (D) $12x(x + 3)^2(x - 3)$
   (E) $24x(x + 3)(x - 3)$

19. Solve for x: $x^2 = 2x + 2$.

   (A) $1 \pm 2\sqrt{3}$
   (B) $1 \pm \sqrt{3}$
   (C) $2 \pm 2\sqrt{3}$
   (D) $-2 \pm 2\sqrt{3}$
   (E) There are no real solutions.

20. Simplify $\left(\dfrac{-2xy^{-3}}{z^{-2}}\right)^{-4}$

(A) $\dfrac{8y^{12}}{x^4 z^8}$

(B) $\dfrac{y^{12} z^8}{8x^4}$

(C) $\dfrac{-y^{12} z^8}{8x^4}$

(D) $\dfrac{y^{12}}{16x^4 z^8}$

(E) $\dfrac{-y^{12}}{16x^4 z^8}$

21. What is the sum of the first 50 terms of the sequence $\{-49, -45, -41, -37, \ldots\}$?

☐

22. Two functions $f(x)$ and $g(x)$ are defined at $x = 3$ by the table below. $f(x)$ is an even function and $g(x)$ is an odd function. A third function $h(x)$ is defined by $f(x) + g(x)$. What is the value of $h(-3)$?

x	$f(x)$	$g(x)$
3	4	-5

(A) -1
(B) 9
(C) 1
(D) -9
(E) not enough information

23. In order to solve the equation $\sqrt{9x+4} - 2x = 1$, you need to solve which of the following equations?

(A) $3x + 2 = 2x + 1$
(B) $4x^2 - 9x - 3 = 0$
(C) $4x^2 - 5x - 3 = 0$
(D) $77x^2 - 4x + 15 = 0$
(E) $4x^2 + x - 1 = 0$

24. $5x(x - 4x^2) - (2x^2 + x^3 - 1) - (x + 2) =$

   (A) $-19x^3 + 3x^2 - x + 1$
   (B) $-21x^3 + 3x^2 - x - 1$
   (C) $-19x^3 + 3x^2 - x - 3$
   (D) $-21x^3 + 3x^2 - x + 3$
   (E) $-21x^3 + 3x^2 - x - 3$

25. For all values of x for which the expression is defined, $\dfrac{\frac{x^2 - 4x - 12}{2x + 2}}{\frac{x - 6}{x^2 - 1}} =$

   (A) $\dfrac{x + 2}{2x - 2}$
   (B) $x^2 + x - 1$
   (C) $\dfrac{x^2 + x - 2}{2}$
   (D) $x^2 + 3x + 1$
   (E) $\dfrac{x^2 + 3x + 2}{2}$

26. Which of the following is a factor of $x^3 + 1000$?

   (A) $x^2 + 100$
   (B) $x^2 - 100$
   (C) $x^2 - 10x + 100$
   (D) $x^2 + 10x + 100$
   (E) $x^2 - 20x + 100$

27. Which value of a makes the following equation true?

   $\begin{vmatrix} a & -6 \\ -2 & a \end{vmatrix} - \begin{vmatrix} 1 & a \\ 2 & a \end{vmatrix} = 8$

   I. 4
   II. -5
   III. 5

   (A) I only
   (B) II only
   (C) III only
   (D) I and II only
   (E) I and III only

28. Which of the following expressions are irrational?

   I. $\sqrt{81} - \sqrt{80}$
   II. $\left(\sqrt{18} - \sqrt{2}\right)^2$
   III. $4.23 - 4.\overline{23}$

   (A) I only
   (B) II only
   (C) III only
   (D) I and II only
   (E) I and III only

29. The graph below is described by which of the following inequalities?

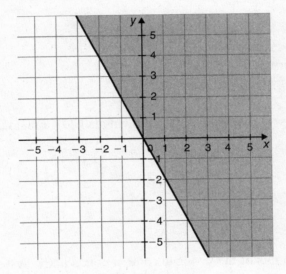

   (A) $2x - y > 0$
   (B) $2x - y \geq 0$
   (C) $2x + y > 0$
   (D) $2x + y \geq 0$
   (E) $x + 2y \geq 0$

30. Which of the following statements are true?

   I. When $P(x) = x^4 - 2x^2 - x + 1$ is divided by $x - 2$, the remainder is 7.
   II. The cubic $P(x) = (x - 3)(x^2 + 2x + 1)$ has three distinct roots.
   III. The polynomial $P(x) = x^4 + ax - b$, where a and b are real numbers, must have at least one root.

   (A) I only
   (B) II only
   (C) III only
   (D) I and III only
   (E) None of the statements are true.

31. Solve for x: $|1 - 2x| \leq 7$.

   (A) $-3 \leq x \leq 4$
   (B) $-4 \leq x \leq 3$
   (C) $x \geq 3$ or $x \leq -4$
   (D) $x \geq 4$ or $x \leq -3$
   (E) $0 \leq x \leq 4$

32. Suppose $a = -\dfrac{1}{2}$. Which of the following is NOT a real number?

   (A) a
   (B) a^2
   (C) a^{-1}
   (D) $a^{1/2}$
   (E) 2^a

33. Which quadrants of the xy-plane contain points in the system $\begin{cases} y < 2x \\ x > 2 \end{cases}$?

   (A) I
   (B) I, IV
   (C) I, II, III
   (D) II, III
   (E) II, III, IV

34. Let $z = 2 + i$. If $\bar{z}$ is the complex conjugate of z, which of the following points shown on the figure below represents $z \cdot \bar{z}$ graphed on the complex plane?

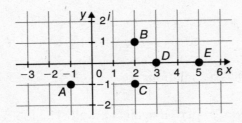

(A) A
(B) B
(C) C
(D) D
(E) E

35. $\dfrac{x-3}{x-2} - \dfrac{x-4}{x+3} =$

(A) $\dfrac{-1-6x}{(x-2)(x+3)}$

(B) $\dfrac{6x-17}{(x-2)(x+3)}$

(C) $\dfrac{1}{(x-2)(x+3)}$

(D) $\dfrac{-17}{(x-2)(x+3)}$

(E) $\dfrac{-1}{5}$

36. Given the system of equations, $\begin{cases} x - y = 1 \\ x^2 - y^2 = 7 \end{cases}$, what value of x will be a solution of the system?

   (A) −4
   (B) 4
   (C) −3
   (D) 3
   (E) The system has no solutions.

37. A store has oranges displayed against a wall. The top four rows of the display of oranges looks like the figure below. If there are 25 rows, what is the total number of oranges in the bottom four rows?

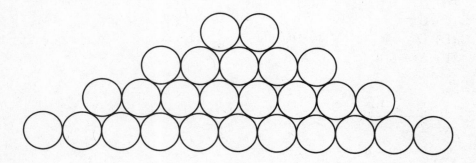

   (A) 74
   (B) 272
   (C) 278
   (D) 290
   (E) 950

38. The following series of algebraic steps simplifies the expression
   $4[5 + (-2 + 3) + 2]$. It uses three algebraic laws: associative, commutative, and distributive. What is the order of the laws that are used as the expression is simplified in the following steps?

   $4[5 + (-2 + 3) + 2]$
   $= 4[5 + (3 - 2) + 2]$
   $= 4[5 + 3 + (-2 + 2)]$
   $= 4(5) + 4(3)$
   $= 20 + 12$
   $= 32$

(A) Distributive, Associative, Commutative
(B) Associative, Distributive, Commutative
(C) Commutative, Distributive, Associative
(D) Associative, Commutative, Distributive
(E) Commutative, Associative, Distributive

39. Which of the following inequalities could give the graph shown in the figure below?

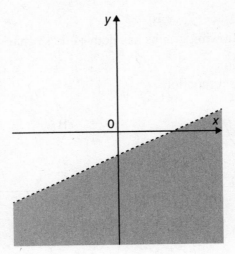

(A) $3y < 2x$
(B) $2x + 3y > 6$
(C) $2x - 3y > 6$
(D) $2x + 3y < 6$
(E) $2x - 3y < 6$

40. Six people start a rumor at 9:00 a.m. If the number of people knowing the rumor doubles every 15 minutes (they spread it over the Internet), how many people know the rumor by 11:15 a.m.?

☐

41. If the equation $9^{3x-1} = \left(\dfrac{1}{3}\right)^{1-2x}$ is solved for x, which of the following is the largest?

(A) x
(B) $\dfrac{1}{x}$
(C) x^2
(D) x^3
(E) $\sqrt{x}$

42. Which of the following graphs has both of these characteristics?

(I) It is a function.
(II) Its inverse is a function.

(A)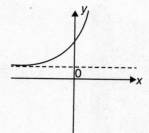

(B)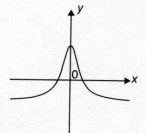

(C)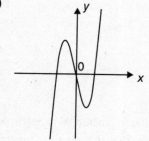

(D)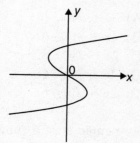

(E)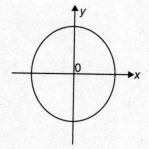

43. What is the third term in the expansion of $\left(2x^2 - \dfrac{y}{2}\right)^6$?

   (A) $240x^8 y^2$
   (B) $20x^6 y^3$
   (C) $-20x^6 y^3$
   (D) $-60x^8 y^2$
   (E) $60x^8 y^2$

44. Which of the following lines has the steepest slope?

   (A) $y = 1$
   (B) $2x + y = 1$
   (C) $x + 2y = 1$
   (D) $3x - 2y = 1$
   (E) $2x + 3y = 1$

45. What is the simplified expression for $\dfrac{(a-b)^2}{a^2-b^2} \div \dfrac{a^2+b^2}{(a+b)^2}$?

   (A) $\dfrac{a+b}{a-b}$

   (B) $\dfrac{a-b}{a+b}$

   (C) $\dfrac{a^2+b^2}{a^2-b^2}$

   (D) $\dfrac{a^2-b^2}{a^2+b^2}$

   (E) $a - b$

46. For how many integer values of x is $|x - 3| + |x + 3| = 6$?

   (A) 0
   (B) 3
   (C) 5
   (D) 7
   (E) 9

47. What is the x-intercept of the graph of the equation $\dfrac{7y}{2} - \dfrac{2x}{5} = 4$?

(A) 10
(B) −10
(C) $\dfrac{7}{8}$
(D) $\dfrac{8}{7}$
(E) $-\dfrac{8}{7}$

48. If $m = 2$ and $n = \dfrac{1}{2}$, arrange the following expressions in order from largest to smallest.

   I. $10^m \cdot 10^n$
   II. $(10^m)^n$
   III. $\dfrac{10^m}{10^n}$

(A) III, II, I
(B) II, III, I
(C) II, I, III
(D) I, III, II
(E) I, II, III

49. Find the value of $\dfrac{8^{3/2}}{2^{-1/2}}$.

(A) 32
(B) 16
(C) 8
(D) $8\sqrt{2}$
(E) $4\sqrt{2}$

50. The equation of the graph in the figure below is given by $y = x^2 + 3x + c$. What is a possible value for c?

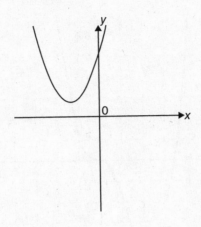

(A) 3
(B) 2
(C) 1
(D) 0
(E) −1

51. If $f(x) = \dfrac{3}{2}x - 8$ and $f^{-1}(x)$ denotes the inverse function of f, which of the following is an expression for $f^{-1}(x)$?

(A) $y = \dfrac{2x + 16}{3}$

(B) $y = \dfrac{3}{2}x + 8$

(C) $y = \dfrac{2}{3}x + 8$

(D) $y = -\dfrac{3}{2}x + 8$

(E) $y = \dfrac{2x + 8}{3}$

52. In the figure below, $f(x) = x^3$, and $g(x)$ is a transformation of $f(x)$. Which of the following is the equation of g?

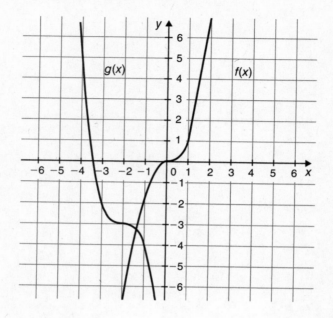

(A) $g(x) = -3x^3 - 3$
(B) $g(x) = -(x + 3)^3 - 2$
(C) $g(x) = -(x - 3)^3 - 2$
(D) $g(x) = -(x + 2)^3 - 3$
(E) $g(x) = -(x - 2)^3 - 3$

53. What is the range of the function $y = |x - 3| + 1$?

(A) $y \geq 1$
(B) $y \geq 3$
(C) $1 \leq y \leq 3$
(D) $-1 \leq y \leq 3$
(E) $y \leq 1$ or $y \geq 3$

54. Which of the following monomials has the highest degree?

(A) 100
(B) $50x$
(C) $10x^5$
(D) $-x^4 y^4$
(E) $-4x^3 y^2 z$

55. If $\log_8(2x-4) = \dfrac{1}{2}$, what is the value of x?

   (A) 0
   (B) 4
   (C) $\sqrt{2} + 2$
   (D) $2\sqrt{2} + 2$
   (E) $2\sqrt{2}$

56. A sequence is given by $\{-2, 4, -8, 16, ...\}$. Which of the following is a formula for the nth term of the sequence?

   (A) $a_n = (-1)2^n$
   (B) $a_n = (-2)^n$
   (C) $a_n = (-2)^{n+1}$
   (D) $a_n = (-1)(-2)^n$
   (E) $a_n = 2 - 2(n-1)$

57. Greg goes into a pet shop with $10. He wants to buy fish for his aquarium. Mollies cost $2.50 and tetras cost $1.50 each. If he buys as many fish as possible with his money and buys at least one of each type of fish, what is the difference between the maximum number of fish he can buy and the minimum?

   (A) 0
   (B) 1
   (C) 2
   (D) 3
   (E) 4

58. Simplify the expression $\dfrac{\dfrac{1}{x}+\dfrac{2}{3}}{\dfrac{1}{x}-\dfrac{3}{4}}$.

(A) $\dfrac{12+8x}{12-9x}$

(B) $1-\dfrac{8}{9}x$

(C) $\dfrac{1}{x}-\dfrac{8}{9}x$

(D) $-\dfrac{8}{9}$

(E) $-\dfrac{8}{9}x$

59. Eliminating fractions in the equation $\dfrac{1}{x+y}-\dfrac{1}{x-y}=1$ leads to which of the following equations?

(A) $x^2 - y^2 = 0$
(B) $x^2 - 2x + 2y - y^2 = 0$
(C) $x^2 + 2y - y^2 = 0$
(D) $x^2 - 2y - y^2 = 0$
(E) $x^2 - 2x - y^2 = 0$

60. Using the figure below, find the equation of the line perpendicular to the given line and passing through the given point.

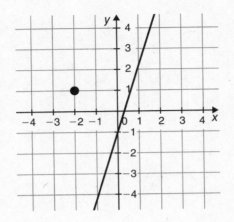

(A) $3y - x = 5$
(B) $3x + y = -5$
(C) $y - 3x = 7$
(D) $2x = 3y$
(E) $x + 3y = 1$

PRACTICE TEST 1

Answer Key

	Answer	Topic	Type
1.	E	Number Systems	Routine
2.	A	Equations/Inequalities	Routine
3.	D	Functions	Nonroutine
4.	E	Equations/Inequalities	Nonroutine
5.	A	Equations/Inequalities	Routine
6.	4	Number Systems	Routine
7.	C	Equations/Inequalities	Routine
8.	B	Functions	Nonroutine
9.	B	Equations/Inequalities	Nonroutine
10.	A	Algebraic Operations	Nonroutine
11.	D	Functions	Routine
12.	7	Functions	Nonroutine
13.	E	Functions	Nonroutine
14.	D	Number Systems	Nonroutine
15.	D	Functions	Routine
16.	A	Number Systems	Routine
17.	B	Equations/Inequalities	Nonroutine
18.	C	Algebraic Operations	Nonroutine
19.	B	Equations/Inequalities	Routine
20.	D	Algebraic Operations	Routine
21.	2450	Number Systems	Routine
22.	B	Functions	Nonroutine
23.	C	Equations/Inequalities	Routine
24.	B	Algebraic Operations	Routine
25.	C	Algebraic Operations	Routine
26.	C	Algebraic Operations	Nonroutine
27.	D	Number Systems	Nonroutine
28.	A	Number Systems	Nonroutine
29.	D	Equations/Inequalities	Routine
30.	A	Functions	Nonroutine

	Answer	Topic	Type
31.	A	Equations/Inequalities	Routine
32.	D	Algebraic Operations	Nonroutine
33.	B	Equations/Inequalities	Routine
34.	E	Number Systems	Routine
35.	B	Algebraic Operations	Routine
36.	B	Equations/Inequalities	Nonroutine
37.	C	Number Systems	Nonroutine
38.	E	Algebraic Operations	Nonroutine
39.	C	Equations/Inequalities	Nonroutine
40.	3072	Number Systems	Nonroutine
41.	B	Equations/Inequalities	Nonroutine
42.	A	Functions	Routine
43.	E	Number Systems	Routine
44.	B	Functions	Nonroutine
45.	D	Algebraic Operations	Routine
46.	D	Equations/Inequalities	Nonroutine
47.	B	Functions	Routine
48.	D	Algebraic Operations	Routine
49.	A	Algebraic Operations	Nonroutine
50.	A	Functions	Nonroutine
51.	A	Functions	Routine
52.	D	Functions	Routine
53.	A	Functions	Nonroutine
54.	D	Algebraic Operations	Routine
55.	C	Equations/Inequalities	Routine
56.	B	Number Systems	Nonroutine
57.	C	Functions	Nonroutine
58.	A	Algebraic Operations	Routine
59.	C	Algebraic Operations	Nonroutine
60.	E	Functions	Routine

PRACTICE TEST 1

Detailed Explanations of Answers

1. **(E)** 4.563563... is a repeating decimal so it is rational. If it is rational, it is real. And if it is real, it is also complex.

2. **(A)**

    $-9 \leq -4x - 3 < 1$

    $-6 \leq -4x < 4$

    $\dfrac{6}{4} \geq x > -1$

    $\dfrac{3}{2} \geq x > -1$

    $(-1, \dfrac{3}{2}]$

3. **(D)** For $f(x)$, $x \neq 4$

    $f(g(x)) = \dfrac{1}{x^2 - 4}$ and $x^2 - 4 = 0$ when $x = \pm 2$.

    So the domain is all real values except $2, -2, 4$.

4. **(E)**

    (A) is a linear function that must have one root $\left(x = \dfrac{5}{4}\right)$.

    (B) is a quadratic function that has one root ($x = -3$).

    (C) is a quadratic function that has no real roots.

    (D) is a cubic function with only one real root ($x = 1$).

    (E) is a quadratic function. The discriminant $= 3^2 - 4(1)(1) = 5$, which is positive. So the equation has two real roots :

    $\dfrac{-3 - \sqrt{5}}{2}$ and $\dfrac{-3 + \sqrt{5}}{2}$.

5. **(A)**

$(2x-1)-3x=9$
$2x-1-3x=9$
$-x=10$
$x=-10$

Check: $|-20-1|+30\stackrel{?}{=}9$
$|-21|+30\stackrel{?}{=}9$
$21+30\stackrel{?}{=}9$
Does not check

$-(2x-1)-3x=9$
$-2x+1-3x=9$
$-5x=8$
$x=\dfrac{-8}{5}$

Check: $\left|-\dfrac{16}{5}-1\right|+\dfrac{24}{5}\stackrel{?}{=}9$
$\dfrac{21}{5}+\dfrac{24}{5}\stackrel{?}{=}9$
$\dfrac{45}{5}=9$
Checks

6. **(4)**

$\sum\limits_{n=0}^{5}(2+x^2-2^x)$

$= (2 + 0 - 1) + (2 + 1 - 2) + (2 + 4 - 4) + (2 + 9 - 8) +$
$(2 + 16 - 16) + (2 + 25 - 32)$

$= 1 + 1 + 2 + 3 + 2 - 5 = 4$

7. **(C)**

$\log_2 8 = x$ $\log_8 2 = y$
$2^x = 8$ $8^y = 2$
$2^x = 2^3$ $2^{3y} = 2^1$
$x = 3$ $3y = 1$
                  $y = \dfrac{1}{3}$

$3+\dfrac{1}{3}=\dfrac{9}{3}+\dfrac{1}{3}=\dfrac{10}{3}$

8. **(B)**

x	y
-2	$-8 - 20 - 12 + 1 = -39$
-1	$-1 - 5 - 6 + 1 = -11$
0	$0 - 0 - 0 + 1 = 1$
1	$1 - 5 + 6 + 1 = 3$
2	$8 - 20 + 12 + 1 = 1$
3	$27 - 45 + 18 + 1 = 1$

There is a sign change in y between $x = -1$ and $x = 0$, so there must be a root in that interval.

9. **(B)** The graph of $y = (x - 1)^2$ is the parabola $y = x^2$ shifted 1 unit to the right.

The graph of $y = (x + 1)^2$ is the parabola $y = x^2$ shifted 1 unit to the left.

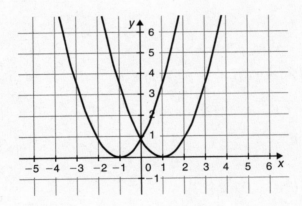

It has one point of intersection, $(0, 1)$.

Alternatively, use the fact that the equations of the curves are equal at the point of intersection.

$x^2 - 2x + 1 = x^2 + 2x + 1$

$4x = 0$

So $x = 0, y = 1$ is the only common point.

10. **(A)**

$$\frac{x^{-2}}{y^{-1}} = \frac{y}{x^2}$$

The multiplicative inverse to $\frac{y}{x^2}$ is the expression by which we multiply $\frac{y}{x^2}$ to get 1. This expression is $\frac{x^2}{y}$.

11. **(D)**

$$m = \frac{2-(-1)}{4-2} = \frac{2+1}{2} = \frac{3}{2}$$

$y - y_1 = m(x - x_1)$ or $y - y_2 = m(x - x_2)$

$y + 1 = \frac{3}{2}(x - 2)$ $y - 2 = \frac{3}{2}(x - 4)$

$2y + 2 = 3x - 6$ $2y - 4 = 3x - 12$

$3x - 2y = 8$ $3x - 2y = 8$

12. **(7)**

$g(5) = |2 - 5| = |-3| = 3$.

Then $f(g(5)) = f(3) = 3^2 - (4)(3) + 10 = 9 - 12 + 10 = 7$.

13. **(E)** For $\sqrt{x+2}$, $x \geq -2$, and for $x + 1 \neq 0$, $x \neq -1$.

The domain is $x \geq -2$, $x \neq -1$ or $[-2, -1) \cup (-1, \infty)$.

14. **(D)**

(A) $4! = 4(3)(2)(1) = 24$

(B) $\frac{6!}{4!} = \frac{(6)(5)(4)(3)(2)(1)}{(4)(3)(2)(1)} = 6(5) = 30$

(C) $\frac{7!}{5!\,2!} = \frac{(7)(6)}{(2)(1)} = \frac{42}{2} = 21$

(D) $\frac{8!}{4!\,4!\,2!} = \frac{(8)(7)(6)(5)}{(4)(3)(2)(1)(2)(1)} = 7(5) = 35$

(E) $\frac{20!}{19!} = \frac{20(19)(18)\ldots(1)}{(19)(18)\ldots(1)} = 20$

PRACTICE TEST 1: DETAILED EXPLANATIONS OF ANSWERS

15. **(D)** Applying the vertical line test, (D) is the only curve for which a vertical line will intersect the graph in more than one location.

16. **(A)** Multiplying top and bottom by the complex conjugate yields
$$\left(\frac{3+i}{1-i}\right)\left(\frac{1+i}{1+i}\right) = \frac{3+3i+i+i^2}{1-i^2} = \frac{3+4i+(-1)}{1-(-1)} = \frac{2+4i}{2} = 1+2i$$

17. **(B)**
$\log_8 64 = y \Rightarrow 8^y = 64$ and $y = 2$

So $4^{x-1} = 2$

$(2^2)^{x-1} = 2$

$2^{2x-2} = 2^1$

$2x - 2 = 1$

$2x = 3$

$x = \frac{3}{2}$

18. **(C)**
$$\frac{x}{2x+6} - \frac{3}{3x-9} + \frac{5}{4x^2+12x} = \frac{x}{2(x+3)} - \frac{3}{3(x-3)} + \frac{5}{4x(x+3)}$$

LCD = $12x(x+3)(x-3)$

19. **(B)**
$x^2 - 2x - 2 = 0$

$x = \dfrac{2 \pm \sqrt{(-2)^2 - 4(1)(-2)}}{2} = \dfrac{2 \pm \sqrt{4+8}}{2}$

$x = \dfrac{2 \pm \sqrt{12}}{2} = \dfrac{2 \pm 2\sqrt{3}}{2} = 1 \pm \sqrt{3}$

20. **(D)**
$$\left(\frac{-2xy^{-3}}{z^{-2}}\right)^{-4} = \frac{(-2)^{-4} x^{-4} y^{12}}{z^8} = \frac{y^{12}}{(-2)^4 x^4 z^8} = \frac{y^{12}}{16 x^4 z^8}$$

21. **(2450)**

This is an arithmetic sequence using the formula $a_n = a_1 + (n-1)d$, where $a_1 = -49$, $d = 4$, and $n = 50$. So $a_{50} = -49 + 49(4) = 147$. The formula for the sum of the first n terms is given by the formula $S_n = \dfrac{n}{2}(a_1 + a_n)$. Thus, for $n = 50$, $S_{50} = \dfrac{50}{2}(-49 + 147) = 2450$.

22. **(B)**

Since f is even, f is symmetric to the y-axis, and $f(3) = f(-3) = 4$.

Since g is odd, g is symmetric to the origin, and $g(-3) = 5$.

$h(-3) = f(-3) + g(-3) = 4 + 5 = 9$

23. **(C)**

$\sqrt{9x+4} = 2x+1$

$9x + 4 = (2x + 1)^2$

$9x + 4 = 4x^2 + 4x + 1$

$0 = 4x^2 - 5x - 3$

24. **(B)**

$5x(x - 4x^2) - (2x^2 + x^3 - 1) - (x + 2)$

$= 5x^2 - 20x^3 - 2x^2 - x^3 + 1 - x - 2$

$= -21x^3 + 3x^2 - x - 1$

25. **(C)**

$\dfrac{\dfrac{x^2 - 4x - 12}{2x + 2}}{\dfrac{x - 6}{x^2 - 1}} = \dfrac{(x-6)(x+2)}{2(x+1)} \cdot \dfrac{(x+1)(x-1)}{(x-6)}$

$\dfrac{(x+2)(x-1)}{2} = \dfrac{x^2 + x - 2}{2}$

26. **(C)**

$x^3 + 1000 = x^3 + 10^3$

Use the sum of cubes formula to get $(x + 10)(x^2 - 10x + 100)$.

PRACTICE TEST 1: DETAILED EXPLANATIONS OF ANSWERS | 279

27. (D)

$a^2 - 12 - (a - 2a) = 8$

$a^2 + a - 20 = 0$

$(a - 4)(a + 5) = 0$

$a = 4, a = -5$

Note that this problem can be done by trial-and-error.

$a = 4: 16 - 12 - (4 - 8) = 4 + 4 = 8$

$a = -5: 25 - 12 - (-5 + 10) = 13 - 5 = 8$

$a = 5: 25 - 12 - (5 - 10) = 13 + 5 = 18 \neq 8$

28. (A)

I. $\sqrt{81} - \sqrt{80} = 9 - 4\sqrt{5}$, which is irrational.

II. $\left(\sqrt{18} - \sqrt{2}\right)^2 = 18 - 2\sqrt{36} + 2 = 18 - 12 + 2 = 8$, which is rational.

III. $4.23 - 4.\overline{23}$ is rational because $4.\overline{23}$ is a repeating decimal, which is always rational.

29. (D) The solid line is $y = -2x$ and the shading is above, so the inequality is $y \geq -2x$.

The graph is $2x + y \geq 0$.

30. (A)

I. Finding the remainder when $P(x)$ is divided by $x - 2$ is the same as finding $P(2)$:

$P(2) = 16 - 2(4) - 2 + 1 = 16 - 8 - 2 + 1 = 7$. True.

II. $P(x) = (x - 3)(x^2 + 2x + 1) = (x - 3)(x + 1)(x + 1)$ has only two distinct roots, $x = 3$ and $x = -1$. False.

III. Since the degree of $P(x)$ is 4, it may have no real roots. This is true for any polynomial functions whose degree is an even number. False.

31. (A)

$(1 - 2x) \leq 7$ $-(1 - 2x) \leq 7$

$1 - 2x \leq 7$ $-1 + 2x \leq 7$

$-6 \leq 2x$ $2x \leq 8$

$-3 \leq x$ $x \leq 4$

Both inequalities check out, so the solution is $-3 \leq x \leq 4$.

32. **(D)**

(A) $\dfrac{-1}{2}$ is real.

(B) $a^2 = \dfrac{1}{4}$, which is real.

(C) $a^{-1} = \dfrac{1}{a} = \dfrac{1}{\frac{-1}{2}} = -2$, which is real.

(D) $a^{1/2} = \sqrt{\dfrac{-1}{2}}$, which is not real.

(E) $2^a = 2^{-1/2} = \dfrac{1}{2^{1/2}} = \dfrac{1}{\sqrt{2}}$, which is real.

33. **(B)** The graph contains all points below the line $y = 2x$, which are in quadrants I, III, and IV, and all points to the right of $x = 2$, which are in quadrants I and IV. So all points in the solution are in quadrants I and IV. The graph is shown below.

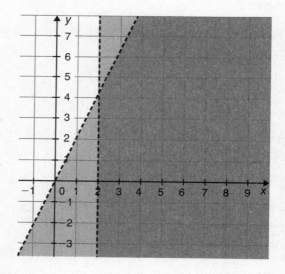

34. **(E)**

$\bar{z} = 2 - i$

$z \cdot \bar{z} = (2 + i)(2 - i) = 4 - i^2 = 4 - (-1) = 5$

$5 = 5 + 0i$

PRACTICE TEST 1: DETAILED EXPLANATIONS OF ANSWERS | 281

35. **(B)**

$$LCD = (x-2)(x+3)$$

$$\left(\frac{x-3}{x-2}\right)\left(\frac{x+3}{x+3}\right) - \left(\frac{x-4}{x+3}\right)\left(\frac{x-2}{x-2}\right)$$

$$= \frac{x^2 - 9 - (x^2 - 6x + 8)}{(x-2)(x+3)}$$

$$= \frac{x^2 - 9 - x^2 + 6x - 8}{(x-2)(x+3)}$$

$$= \frac{6x - 17}{(x-2)(x+3)}$$

36. **(B)** The problem can be done directly by substitution: $y = x - 1$.

$x^2 - (x-1)^2 = 7$

$x^2 - (x^2 - 2x + 1) = 7$

$2x - 1 = 7$

$2x = 8$

$x = 4$

The problem can also be done by trial and error.

(A) if $x = -4$, then equation 1: $-4 - y = 1$ and $y = -5$.

Equation 2: $(-4)^2 - (-5)^2 = 16 - 25 = -9 \neq 7$.

(B) if $x = 4$, then equation 1: $4 - y = 1$ and $y = 3$.

Equation 2: $4^2 - 3^2 = 16 - 9 = 7$.

Stop after you get (B) as the correct solution.

37. **(C)** The sequence is $\{2, 5, 8, 11, \ldots\}$, which is an arithmetic sequence.

Use the formula $a_n = a_1 + (n-1)d$, where $a_1 = 2$, $d = 3$, and $n = 25$; then $a_{25} = 2 + (25 - 1)(3) = 2 + 24(3) = 2 + 72 = 74$.

The 25th row has 74 oranges, so the rows above that have 71, 68, and 65 oranges, respectively. $74 + 71 + 68 + 65 = 278$.

38. **(E)** The first step changes $(-2+3)$ to $(3-2)$, which is the commutative law.

The second step changes $(3-2)+2$ to $3+(-2+2)$, which is the associative law.

The third step changes $4(5+3) = 4(5) + 4(3)$, which is the distributive law.

39. **(C)** The line has a positive slope, a negative y-intercept, and the shading is below. It is best to solve each inequality for y.

(A) $y < \frac{2}{3}x$: positive slope and the shading below, but this would pass through the origin, so it can't be right.

(B) $y > -\frac{2}{3}x + 2$: negative slope and shading above, so it can't be right.

(C) $y < \frac{2}{3}x - 2$: positive slope, shading below, and a negative y-intercept, so it may be right.

(D) $y < -\frac{2}{3}x + 2$: negative slope, and a positive y-intercept so it can't be right.

(E) $y > \frac{2}{3}x - 2$: positive slope but shading above, so it can't be right.

40. **(3072)** This is a geometric growth sequence: 6, 12, 24, ... The formula for geometric growth is $a_n = a_1 r^{n-1}$; where $a_1 = 6$ and $r = 2$. In order to determine the value of n, count 9:00 am. as the time of the first number. Then the time from 9:00 a.m. to 11:15 a.m. represents 2.25 hours, which is 135 minutes. Since $135 \div 15 = 9$, there are nine more numbers. The answer is thus the tenth term of the sequence 6, 12, 24, ... Thus, $a_{10} = a_1 r^{10-1} = (6)(2^{10-1}) = 6(2)^9 = 6(512) = 3072$.

This solution could also be found by simply doubling each term: 6, 12, 24, 48, 96, 192, 384, 768, 1536, 3072.

41. **(B)**

$(3^2)^{3x-1} = (3^{-1})^{1-2x}$

$3^{6x-2} = 3^{-1+2x}$

$6x - 2 = -1 + 2x$

$4x = 1$

$x = \dfrac{1}{4}$

(A) $x = \dfrac{1}{4}$

(B) $\dfrac{1}{x} = 4$

All of the others values will obviously be fractions, so (B) is the correct answer.

42. **(A)** To be a function, the graph must pass the vertical line test and for the inverse to be a function, the graph must also pass the horizontal line test, similar to the vertical line test.

(A) passes both

(B) passes the vertical line test but not the horizontal line test

(C) passes the vertical line test but not the horizontal line test

(D) passes the horizontal line test but not the vertical line test

(E) passes neither

43. **(E)** Since the term number is odd, the answer is positive.

$_6C_4 (2x^2)^4 \left(\dfrac{y}{2}\right)^2$

$= \dfrac{6!}{4!\,2!}(16x^8)\left(\dfrac{y^2}{4}\right)$

$= \dfrac{6(5)}{(2)(1)}(16x^8)\left(\dfrac{y^2}{4}\right)$

$= 15(16x^8)\left(\dfrac{y^2}{4}\right)$

$= 60x^8 y^2$

44. **(B)** Use $y = mx + b$, and compare $|m|$ values for the steepness.

(A) $y = 1$ has slope 0; steepness $= 0$.

(B) $y = -2x + 1$ has slope -2: steepness $= 2$.

(C) $y = \dfrac{-x}{2} + \dfrac{1}{2}$ has slope $-\dfrac{1}{2}$: steepness $= \dfrac{1}{2}$.

(D) $y = \dfrac{3x}{2} - \dfrac{1}{2}$ has slope $\dfrac{3}{2}$: steepness $= \dfrac{3}{2}$.

(E) $y = \dfrac{-2x}{3} + \dfrac{1}{3}$ has slope $\dfrac{-2}{3}$: steepness $= \dfrac{2}{3}$.

Of these five values for steepness, 2 is the largest.

45. **(D)**

$$\dfrac{(a-b)^2}{a^2-b^2} \div \dfrac{a^2+b^2}{(a+b)^2}$$

$$= \dfrac{(a-b)^2}{a^2-b^2} \cdot \dfrac{(a+b)^2}{a^2+b^2}$$

$$= \dfrac{\cancel{(a-b)}(a-b)}{\cancel{(a+b)}\cancel{(a-b)}} \cdot \dfrac{\cancel{(a+b)}(a+b)}{a^2+b^2}$$

$$= \dfrac{(a-b)(a+b)}{a^2+b^2}$$

$$= \dfrac{a^2-b^2}{a^2+b^2}$$

46. **(D)** Do this by finding values that satisfy the equation.

$x = 0$: $|-3| + |3| = 6$

$x = 1$: $|-2| + |4| = 6$ $x = -1$: $|-4| + |2| = 6$

$x = 2$: $|-1| + |5| = 6$ $x = -2$: $|-5| + |1| = 6$

$x = 3$: $|0| + |6| = 6$ $x = -3$: $|-6| + |0| = 6$

There are seven integers that satisfy the equation, namely, $-3, -2, -1, 0, 1, 2,$ and 3.

Any values greater than 3 or less than -3 make the expression greater than 6.

PRACTICE TEST 1: DETAILED EXPLANATIONS OF ANSWERS

47. **(B)** At the x-intercept, $y = 0$, so find the value of x when $y = 0$.

$$\frac{7y}{2} - \frac{2x}{5} = 4$$

$$0 - \frac{2x}{5} = 4$$

$$-2x = 20$$

$$x = -10$$

48. **(D)**

  I. $10^m \cdot 10^n = 10^{m+n} = 10^{2.5}$, which is greater than $100\ (=10^2)$.

  II. $(10^m)^n = 10^{mn} = 10^1 = 10$.

  III. $\dfrac{10^m}{10^n} = 10^{m-n} = 10^{1.5}$, which is greater than 10 but less than 100.

49. **(A)** This solution can be done in two ways

$$\frac{8^{3/2}}{2^{-1/2}} = (8^{3/2})(2^{1/2}) = (\sqrt{8})^3(\sqrt{2}) \qquad\qquad \frac{8^{3/2}}{2^{-1/2}} = (8^{3/2})(2^{1/2})$$
$$= (\sqrt{8})(\sqrt{8})(\sqrt{8})(\sqrt{2}) \qquad\qquad\qquad = (2^3)^{3/2}(2^{1/2})$$
$$= (\sqrt{64})(\sqrt{16}) \qquad\text{or}\qquad\qquad = (2^{9/2})(2^{1/2})$$
$$= 8(4) \qquad\qquad\qquad\qquad\qquad = 2^{9/2+1/2} = 2^{10/2} = 2^5$$
$$= 32 \qquad\qquad\qquad\qquad\qquad\quad = 32$$

50. **(A)** This quadratic has no roots, so the discriminant $b^2 - 4ac < 0$. Using $a = 1, b = 3$, we get

$$3^2 - 4c < 0$$
$$9 < 4c$$
$$\frac{9}{4} < c$$

The only choice for $c > \dfrac{9}{4}$ is 3.

The problem could also be done by trial-and-error by testing each choice for $b^2 - 4ac < 0$.

  (A) $9 - 4(3) < 0$ and $-3 < 0$, true.
  (B) $9 - 4(2) < 0$ and $1 < 0$, false.
  (C) $9 - 4(1) < 0$ and $5 < 0$, false.
  (D) $9 - 4(0) < 0$ and $9 < 0$, false.
  (E) $9 - 4(-1) < 0$ and $13 < 0$, false.

51. **(A)** For the inverse of a function, switch x and y and solve for the new y.
Inverse: $x = \dfrac{3}{2}y - 8$
$2x = 3y - 16$
$2x + 16 = 3y$
$\dfrac{2x+16}{3} = y$

52. **(D)** The f curve has been first reflected across the x-axis giving $g(x) = -x^3$.

   It is then shifted 2 units to the left, giving $g(x) = -(x+2)^3$.

   Finally it is shifted 3 units down, giving $g(x) = -(x+2)^3 - 3$.

53. **(A)** Since $|x - 3| \geq 0$, y must be ≥ 1.

   A second approach would be graphically:

   The graph of $y = |x|$ is a V-shaped curve whose range is $y \geq 0$

   The graph of $y = |x - 3|$ shifts the curve 3 units to the right which doesn't affect the range.

   The graph of $y = |x - 3| + 1$ shifts the curve 1 unit up so the range is $y \geq 1$.

   Alternatively, points can be plotted to show that $y \geq 1$:

x	3	4	2	5	1
y	1	2	2	3	3

54. **(D)** The degree is the sum of the exponents.

   (A) has degree 0.

   (B) has degree 1.

   (C) has degree 5.

   (D) has degree 8.

   (E) has degree 6.

55. **(C)**

$8^{1/2} = 2x - 4$

$\sqrt{8} = 2x - 4$

$2\sqrt{2} + 4 = 2x$

$\dfrac{2\sqrt{2} + 4}{2} = x$

$x = \sqrt{2} + 2$

56. **(B)** This is a geometric sequence whose form is $a_n = a_1 r^{n-1}$.

Since $a_1 = -2$ and $r = -2$, the formula is $a_n = (-2)(-2)^{n-1} = (-2)^n$.

This can be done by trial-and-error as well:

(A) $a_n = \{-2, -4, -8, \ldots\}$. Not the formula because $a_2 \neq 4$.

(B) $a_n = \{-2, 4, -8, \ldots\}$. This is it! You don't have to go any further, but if you do, you can eliminate (C) and (D) right away because $a_1 \neq -2$. Choice (E) is eliminated because $a_2 \neq 4$. Don't do more checking than necessary with trial-and-error.

57. **(C)** If x = number of mollies and y = number of tetras, the equation is $2.5x + 1.5y < 10$.

Since x and y must be integers, it is best to set up a chart.

x	Cost of Mollies	y	Cost of Tetras	Total Cost	Total Fish
3	$7.50	1	$1.50	$9.00	4
2	$5.00	3	$4.50	$9.50	5
1	$2.50	5	$7.50	$10.00	6

The difference between the maximum and minimum number of fish he can buy is $6 - 4 = 2$.

58. **(A)**

The LCD is $12x$.

$$\left(\dfrac{\dfrac{1}{x}+\dfrac{2}{3}}{\dfrac{1}{x}-\dfrac{3}{4}}\right)\left(\dfrac{12x}{12x}\right)=\dfrac{12+8x}{12-9x}$$

59. **(C)**

LCD $= (x+y)(x-y)$, so

$$\cancel{(x+y)}(x-y)\dfrac{1}{\cancel{x+y}}-(x+y)\cancel{(x-y)}\dfrac{1}{\cancel{x-y}}=(x+y)(x-y)$$

$x - y - (x+y) = x^2 - y^2$

$x - y - x - y = x^2 - y^2$

$0 = x^2 + 2y - y^2$

60. **(E)** By observation, the slope of the given line is 3. (It goes up 3 while going to the right 1.)

By calculation, choose two points on the line and calculate the slope. For $(0, -1)$ and $(1, 2)$, the slope is $\dfrac{2-(-1)}{1-0}=3$.

The slope of the perpendicular line is the negative reciprocal, or $m = \dfrac{-1}{3}$.

The given point is $(x_1, y_1) = (-2, 1)$.

Then $y - y_1 = m(x - x_1)$

$y - 1 = \dfrac{-1}{3}(x+2)$

$3y - 3 = -x - 2$

$x + 3y = 1$

PRACTICE TEST 2

CLEP College Algebra

Also available at the REA Study Center (*www.rea.com/studycenter*)

This practice test is also offered online at the REA Study Center. Since all CLEP exams are administered on computer, we recommend that you take the online version of this test to simulate test-day conditions and to receive these added benefits:

- **Timed testing conditions** – helps you gauge how much time you can spend on each question
- **Automatic scoring** – find out how you did on the test, instantly
- **On-screen detailed explanations of answers** – gives you the correct answer and explains why the other answer choices are wrong
- **Diagnostic score reports** – pinpoint where you're strongest and where you need to focus your study

PRACTICE TEST 2

CLEP College Algebra

(Answer sheets appear in the back of the book.)

TIME: 90 Minutes
60 Questions

DIRECTIONS: An online scientific calculator will be available for the questions in this test.

Some questions will require you to select from among five choices. For these questions, select the BEST of the choices given.

Some questions will require you to type a numerical answer in the box provided.

Notes:

(1) Unless otherwise specified, the domain of any function f is assumed to be the set of all real numbers x for which $f(x)$ is a real number.

(2) i will be used to denote $\sqrt{-1}$.

(3) Figures that accompany questions are intended to provide information useful in answering the questions. All figures lie in a plane unless otherwise indicated. The figures are drawn as accurately as possible *except* when it is stated in a specific question that the figure is not drawn to scale. Straight lines and smooth curves may appear slightly jagged on the screen.

1. The greatest rational number among the following five choices is:

   (A) $|-5-1|$

   (B) $\left(\dfrac{5}{2}\right)^2$

   (C) $6.\overline{25}$

   (D) $40^{1/2}$

   (E) $(-2\pi)^2$

2. Arrange these numbers from largest to smallest.

   I. $\left(\dfrac{-1}{4}\right)^{-2}$

   II. $(-4)^{-2}$

   III. $4^{-1/2}$

   (A) III, I, II
   (B) II, I, III
   (C) II, III, I
   (D) I, III, II
   (E) I, II, III

3. In the system of equations $\begin{cases} 7x+3y+1=0 \\ 3x-y-11=0 \end{cases}$, the x-value is

   (A) 0
   (B) −5
   (C) 5
   (D) −2
   (E) 2

4. The equation of the graph below could be:

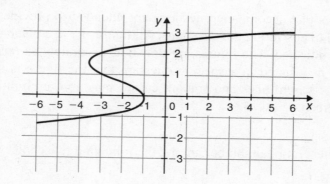

   (A) $y = \dfrac{-x^2}{4} - \dfrac{x}{2} - \dfrac{1}{4}$
   (B) $y = -x^2 + 1$
   (C) $x = y^3 + 2y^2 - 1$
   (D) $x = y^3 - 2y^2 - y - 1$
   (E) $y = x^3$

5. Solve for x: $(3x + 2)(x - 1) = x^2 - 2x - 1$.

   (A) $x = -1$ only

   (B) $x = -1, x = \dfrac{1}{2}$

   (C) $x = -\dfrac{1}{2}$ only

   (D) $x = -\dfrac{1}{2}, x = 1$

   (E) $x = -1, x = -\dfrac{1}{2}$

6. When in expanded form, which of the following is different from the others?

   (A) $(4x + 9)^2$
   (B) $(-4x - 9)^2$
   (C) $|4x + 9|^2$
   (D) $\left(\sqrt{16x^2 + 81}\right)^2$

   (E) All of them are the same.

7. Which of the following is true when comparing these three values?

   I. $\log_5 4 + \log_5 25$
   II. $2\log_5 10$
   III. $\log_5 10 - \log_5 \dfrac{1}{10}$

   (A) All represent different values.
   (B) I and II are the same, III is different.
   (C) I and III are the same, II is different.
   (D) II and III are the same, I is different.
   (E) I, II, and III all represent the same value.

8. Which of the following graphs is NOT a function?

(A)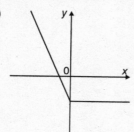

(B)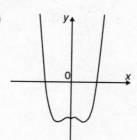

(C)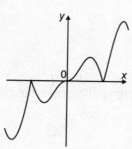

(D)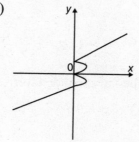

(E)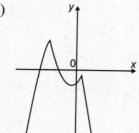

9. Sam wants to multiply (17)(125)(8). He doesn't have a calculator, but realizes that rather than first multiplying 17 and 125 and then multiplying the result by 8, he can multiply 125 by 8 in his head, getting 1,000. And then he multiplies 17 by 1,000 to get 17,000. What mathematical law allows him to do this?

(A) Associative
(B) Closure
(C) Commutative
(D) Distributive
(E) Inverse

10. If $f(x) = x^2 - 1$, find $f(a + 1) - f(a - 1)$

   (A) 0
   (B) $2a^2 + 4a$
   (C) $2a^2 + 4a - 2$
   (D) $4a$
   (E) $4a - 2$

11. What is the third term in the expansion of $\left(2x - \dfrac{1}{4y}\right)^6$?

   (A) $\dfrac{x^4}{y^2}$

   (B) $-\dfrac{15x^4}{y^2}$

   (C) $\dfrac{15x^4}{y^2}$

   (D) $-\dfrac{5x^3}{2y^3}$

   (E) $\dfrac{5x^3}{2y^3}$

12. Two inequalities are solved. The first one has the solution $-3 \leq x < 2$, and the other has the solution $(-1, 4]$. If both inequalities must be true at the same time and x is an integer, how many possible solutions are there?

   (A) 5
   (B) 4
   (C) 3
   (D) 2
   (E) 1

13. Solve for x: $8^{2x-1} = \left(\dfrac{1}{4}\right)^{2-x}$.

    (A) $-\dfrac{1}{4}$

    (B) $\dfrac{1}{8}$

    (C) 1

    (D) -1

    (E) $-\dfrac{7}{4}$

14. If a is the number of solutions to the equation $|3x| = 9$ and b is the number of solutions to the equation $|2x - 5| + 3x = 0$, what is the sum $a + b$?

    (A) 0
    (B) 1
    (C) 2
    (D) 3
    (E) 4

15. Which of the following is equivalent to $\dfrac{-6x^{2/3}}{3x^{-3/2}}$?

    (A) $-2x$

    (B) $\dfrac{-2}{x}$

    (C) $-2x^{13/6}$

    (D) $\dfrac{-2}{x^{5/6}}$

    (E) -2

16. The graphs of $f(x)$ and $g(x)$ are shown below. Find the value of $g(f(-2)) - f(g(-2))$.

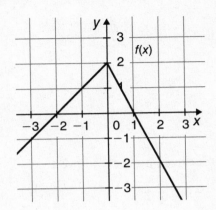
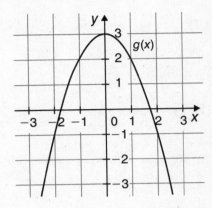

(A) −2
(B) −1
(C) 0
(D) 1
(E) 2

17. A company makes x laptop computers with a cost function given by $C(x) = 400x$. The revenue function when selling x laptops is given by $R(x) = \dfrac{x^2}{2}$. The profit function P is given by $R(x) - C(x)$. Breaking even means the profit is zero. How many laptop computers must the company sell to break even?

18. Which quadratic equation has *at least* one real solution?

   I. $x^2 + 20x + 100 = 0$
   II. $5x^2 + 2x - 16 = 0$
   III. $4x^2 - 7x + 4 = 0$

(A) I only
(B) II only
(C) III only
(D) I and II only
(E) I, II, and III

19. Which of the following functions does NOT include $x = -8$ in its domain?

   (A) $y = -x - 8$
   (B) $y = \dfrac{1}{x^2 + 8}$
   (C) $y = 8^{-x-8}$
   (D) $y = \sqrt{x+8}$
   (E) $y = \dfrac{1}{\sqrt{x^3 - 64x}}$

20. Arrange the following numbers in order from highest to lowest.

   I. $\ln e^2 - \ln e^5$
   II. $2 \ln e^{1/2}$
   III. $\ln 1$

   (A) I, II, III
   (B) I, III, II
   (C) II, III, I
   (D) II, I, III
   (E) III, I, II

21. An inequality that passes through the y-intercept of -1 is graphed in the figure below. What is the inequality?

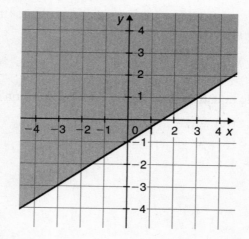

   (A) $3y - 2x - 1 \geq 0$
   (B) $3x - 2y - 2 \leq 0$
   (C) $3x - 2y - 2 \geq 0$
   (D) $2x - 3y - 3 \geq 0$
   (E) $2x - 3y - 3 \leq 0$

22. $(2 - 3i)^2 - (2 - 3i) =$

   (A) $-7 - 9i$
   (B) $-7 - 15i$
   (C) $-7 + 3i$
   (D) $-7 - 3i$
   (E) $7 + 3i$

23. Which of these numbers has the largest value?

   (A) $4^{(3^2)}$
   (B) $4^{(2^3)}$
   (C) 16^6
   (D) $(2^5)^2$
   (E) $\left(\dfrac{1}{4}\right)^{-4^2}$

24. Which of the following quadratic equations has $1 - i$ as a root?

   (A) $x^2 - 4 = 0$
   (B) $x^2 - 2x + 1 = 0$
   (C) $x^2 + 1 = 0$
   (D) $x^2 - 2 = 0$
   (E) $x^2 - 2x + 2 = 0$

25. If $f(x) = \dfrac{3x - 5}{2x + 1}$, find $f^{-1}(x)$.

   (A) $y = \dfrac{x - 5}{2x + 3}$
   (B) $y = \dfrac{2x + 1}{3x - 5}$
   (C) $y = \dfrac{x + 5}{3 - 2x}$
   (D) $y = \dfrac{2}{3}x - \dfrac{1}{5}$
   (E) $y = -\dfrac{2}{3}x + \dfrac{1}{5}$

26. Find the value of $\sum_{n=0}^{3}(-1)^n \frac{1}{2^n}$.

(A) $\frac{15}{8}$

(B) $\frac{5}{8}$

(C) $\frac{7}{12}$

(D) $\frac{-1}{8}$

(E) $\frac{1}{8}$

27. What is the remainder when $x^5 - 10x + 2$ is divided by $x + 2$?

(A) -54
(B) -50
(C) -10
(D) 12
(E) 14

28. Factor $-x^3 + 4x^2 + 4x - 16$.

(A) $-(x - 4)(x + 2)(x - 2)$

(B) $-(x + 2)^2(x - 2)^2$

(C) $-(x + 4)(x + 2)(x - 2)$

(D) $(x - 4)(x + 2)(x - 2)$

(E) $(-x + 2)^2(x - 2)^2$

29. Which of the points in the figure below represents the complex number $\frac{3+i}{i}$ graphed on the complex plane?

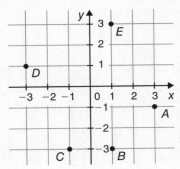

(A) A
(B) B
(C) C
(D) D
(E) E

30. Determine the values of M and N so that the graph of $Mx + Ny = 5$ will contain the points $(3, 1)$ and $(2, -1)$.

(A) $M = -2, N = 11$
(B) $M = -1, N = 8$
(C) $M = 2, N = -1$
(D) $M = 1, N = 2$
(E) $M = 3, N = -1$

31. If $x + y = z$, find the value of $|z - y| + |y - z|$.

(A) $2x$
(B) $2|x|$
(C) $x - z$
(D) $2x - 2z$
(E) 0

32. A theater has 26 rows of seats. There are 16 seats in each of the first two rows, 18 seats in each of the next two rows, 20 seats in each of the next two rows, and so on. Below is a figure showing the first six rows. How many seats are in the theater?

33. Classify the real solutions of $\sqrt{2x+14} - x = 3$.

   (A) two positive solutions
   (B) one positive and one negative solution
   (C) two negative solutions
   (D) one positive solution only
   (E) one negative solution only

34. Which of the following is true?

   I. $144^{-1/2} = \pm \dfrac{1}{12}$

   II. $64^{2/3} = 16$

   III. $(-32)^{-3/5} = \dfrac{1}{8}$

   (A) I only
   (B) II only
   (C) III only
   (D) II and III
   (E) I, II, and III

35. Find the 50th term in the sequence 902, 885, 868, 851, ...

36. Place these three values in order from largest to smallest.

   I. $\log_9 1$

   II. $\log_9 \dfrac{1}{3}$

   III. $\log_{27} 9$

   (A) I, II, III
   (B) II, I, III
   (C) II, III, I
   (D) III, I, II
   (E) III, II, I

37. If $f(x) = \begin{cases} x - x^2, & x \geq 1 \\ -x - 1, & -1 \leq x < 1 \\ 2x + 3, & x < -1 \end{cases}$, find $f(f(f(2)))$.

   (A) -8
   (B) -2
   (C) -1
   (D) 0
   (E) 2

38. Which of the following is a geometric sequence?

   (A) $a_n = \{55, 61, 67, 73, \ldots\}$
   (B) $a_n = \{1, 2, 4, 7, 11, \ldots\}$
   (C) $a_n = \left\{\dfrac{1}{2}, \dfrac{1}{3}, \dfrac{1}{4}, \dfrac{1}{5} \ldots\right\}$
   (D) $a_n = \{\sqrt{2}, \sqrt{4}, \sqrt{8}, \sqrt{12}\ldots\}$
   (E) $a_n = \left\{-12, 3, -\dfrac{3}{4}, \dfrac{3}{16}, \ldots\right\}$

39. Solve for x: $4 \leq 3 - \frac{1}{2}x < \frac{7}{4}$.

(A) $\left[2, \frac{5}{2}\right]$

(B) $\left[2, \frac{5}{2}\right)$

(C) $\left[-2, \frac{5}{2}\right)$

(D) $\left[-2, \frac{5}{2}\right]$

(E) $(-\infty, -2] \cup \left(\frac{5}{2}, \infty\right)$

40. $\dfrac{5}{a+b} - \dfrac{4}{a-b} =$

(A) $\dfrac{a - 9b}{a^2 - b^2}$

(B) $\dfrac{1}{a-b}$

(C) $\dfrac{1}{a+b}$

(D) $\dfrac{1}{a^2 - b^2}$

(E) $\dfrac{1}{2b}$

41. Solve for x: $\log_x (3x + 18) = 2$

(A) $x = 6, x = -3$
(B) $x = 6$ only
(C) $x = -18$ only
(D) $x = \dfrac{82}{3}$ only
(E) $x = 3$ only

42. If two irrational numbers are multiplied together, the result

   I. must be rational
   II. can be rational
   III. can be an integer

   (A) I only
   (B) II only
   (C) III only
   (D) I and III only
   (E) II and III only

43. The graph in the figure below is $f(x)$. It passes through the points $(-1, 5)$ and $(1, 1)$ and has one root. What is a possible equation of the transformation of $f(x)$ that would give it three roots?

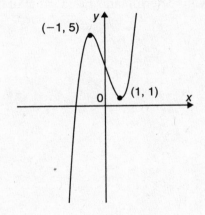

   (A) $f(x - 1)$
   (B) $f(x) - 1$
   (C) $f(x) - 3$
   (D) $f(x + 3)$
   (E) $-f(x)$

44. Solve for x: $\dfrac{5x+2}{6} - \dfrac{3x-1}{9} = \dfrac{2x+3}{12}$

(A) $\dfrac{1}{36}$

(B) $\dfrac{1}{12}$

(C) $\dfrac{-7}{12}$

(D) $\dfrac{-25}{24}$

(E) -1

45. Which of the following determinants has a value of 0?

I. $\begin{vmatrix} -9 & -9 \\ -9 & -9 \end{vmatrix}$

II. $\begin{vmatrix} 9 & -9 \\ -9 & -9 \end{vmatrix}$

III. $\begin{vmatrix} \dfrac{-1}{9} & \dfrac{2}{3} \\ \dfrac{3}{2} & 9 \end{vmatrix}$

(A) I only
(B) II only
(C) III only
(D) I and II only
(E) I, II, and III

46. Do the following division, putting the answer in the most simplified form:
$$\frac{2-x}{x^2+3x} \div \frac{x^2-x-2}{x^2+4x+3}.$$

(A) $\dfrac{2-x}{x^2-2x}$

(B) $\dfrac{2-x}{x-2}$

(C) -1

(D) $-x$

(E) $\dfrac{-1}{x}$

47. The point (2, 3) satisfies the system of equations $\begin{cases} 4x^2+y^2=25 \\ y^2-x^2=5 \end{cases}$. How many additional points satisfy the system?

(A) 0
(B) 1
(C) 2
(D) 3
(E) 4

48. Solve for x: $\dfrac{2x}{5}-\dfrac{5x}{2} \geq 1-x$

(A) $\left[\dfrac{-10}{31},\infty\right)$

(B) $\left(-\infty,\dfrac{10}{31}\right]$

(C) $\left(-\infty,\dfrac{-10}{11}\right]$

(D) $\left(-\infty,\dfrac{10}{11}\right]$

(E) $\left[-\dfrac{10}{11},\infty\right)$

49. A system of inequalities is given by $\begin{cases} y \geq px + 4, \ p > 0 \\ y \geq qx + 4, \ q < 0 \end{cases}$. In what quadrant(s) is the solution?

   (A) I only
   (B) I and II only
   (C) II only
   (D) III and IV only
   (E) I, II, and IV only

50. A function $f(x)$ passes through these points:

x	−4	0	4
f(x)	4	0	−4

   Which of the following could be a graph of the inverse function $f^{-1}(x)$?

   (A)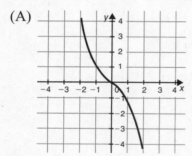

   (B)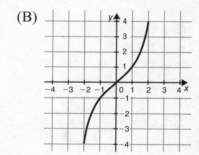

   (C)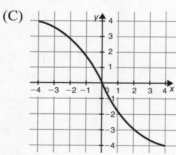

   (D)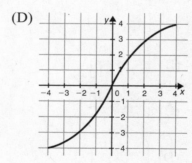

   (E)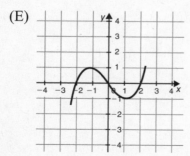

51. When $\dfrac{1+i}{4+2i}$ is expressed in the form of $a + bi$, what is the value of a?

   (A) $\dfrac{1}{2}$

   (B) $\dfrac{3}{10}$

   (C) $\dfrac{1}{4}$

   (D) $\dfrac{1}{6}$

   (E) $\dfrac{1}{10}$

52. A polynomial $y = f(x)$ passes through these points. What is the minimum number of roots it can have?

x	−8	−6	−4	−2	0	2	4	6	8
f(x)	−12	−8	−1	5	3	0	4	−9	1

   (A) 5
   (B) 4
   (C) 3
   (D) 2
   (E) 1

53. If $xy^6 = -192$ and $xy^3 = 24$, find x.

   ☐

54. Let $f(x) = x^2 - x + 2$ and $g(x) = 2 - x$. Let $a = f(g(-2))$, $b = g(f(-2))$, $c = f(g(2))$, and $d = g(f(2))$. Which of the following statements is true?

   (A) $a + b = 0$ only
   (B) $c + d = 0$ only
   (C) $a + b = 0$ and $c + d = 0$
   (D) $a = b$
   (E) $c = d$

55. Simplify $x - [x - x(x - y) + y(x - y)]$

   (A) $(x + y)^2$
   (B) $(x - y)^2$
   (C) $x^2 - y^2$
   (D) $x^2 + y^2$
   (E) 0

56. Line k passes through the points $(-12, 10)$ and $(6, 1)$. Line l is perpendicular to line k with y-intercept 6. Find the x-intercept of line l.

   (A) -9
   (B) -12
   (C) 12
   (D) -3
   (E) 3

57. Solve for x: $|6x - 3| - 15 < 0$.

   (A) $(-2, \infty)$
   (B) $(-\infty, 3)$
   (C) $(-2, 3)$
   (D) $[-2, 3]$
   (E) $(-\infty, -2) \cup (3, \infty)$

58. Simplify $\left(\dfrac{1}{1-\sqrt{2}}\right)^2$.

   (A) $1 + 2\sqrt{2}$
   (B) $1 + \sqrt{2}$
   (C) $2\sqrt{2} - 1$
   (D) $\dfrac{3 - 2\sqrt{2}}{5}$
   (E) $3 + 2\sqrt{2}$

59. A projectile is fired downward from a helicopter 160 feet above the ground. Its height above ground in feet is given by $h(t) = -16t^2 + 48t + 160$. How long in seconds will it take the projectile to strike the ground?

60. A function $y = g(x)$ is given by the graph in the figure below left. If the graph in the figure below right is a transformation of $g(x)$, which of the following equations describes it?

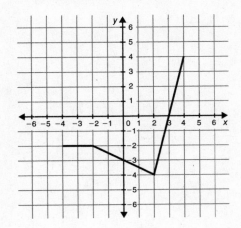
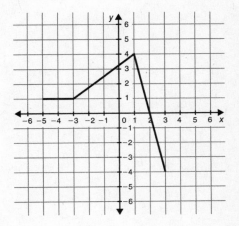

(A) $y = 1 - g(x)$
(B) $y = 1 - g(x + 1)$
(C) $y = 1 - g(x - 1)$
(D) $y = -g(x - 1)$
(E) $y = -g(x + 1)$

PRACTICE TEST 2

Answer Key

	Answer	Topic	Type
1.	C	Number Systems	Routine
2.	D	Algebraic Operations	Routine
3.	E	Equations/Inequalities	Routine
4.	D	Functions	Nonroutine
5.	B	Equations/Inequalities	Routine
6.	D	Algebraic Operations	Nonroutine
7.	E	Equations/Inequalities	Nonroutine
8.	D	Functions	Routine
9.	A	Algebraic Operations	Nonroutine
10.	D	Functions	Routine
11.	C	Number Systems	Nonroutine
12.	D	Algebraic Operations	Nonroutine
13.	A	Equations/Inequalities	Routine
14.	D	Equations/Inequalities	Nonroutine
15.	C	Algebraic Operations	Routine
16.	E	Functions	Nonroutine
17.	800	Functions	Nonroutine
18.	D	Equations/Inequalities	Routine
19.	E	Functions	Routine
20.	C	Equations/Inequalities	Routine
21.	E	Functions	Nonroutine
22.	A	Number Systems	Routine
23.	E	Algebraic Operations	Nonroutine
24.	E	Number Systems	Nonroutine
25.	C	Functions	Routine
26.	B	Number Systems	Routine
27.	C	Functions	Nonroutine
28.	A	Algebraic Operations	Routine
29.	B	Number Systems	Nonroutine
30.	C	Equations/Inequalities	Nonroutine
31.	B	Algebraic Operations	Nonroutine

	Answer	Topic	Type
32.	728	Number Systems	Nonroutine
33.	D	Equations/Inequalities	Nonroutine
34.	B	Algebraic Operations	Routine
35.	69	Number Systems	Routine
36.	D	Equations/Inequalities	Nonroutine
37.	D	Functions	Nonroutine
38.	E	Number Systems	Routine
39.	E	Equations/Inequalities	Routine
40.	A	Algebraic Operations	Routine
41.	B	Equations/Inequalities	Routine
42.	E	Number Systems	Nonroutine
43.	C	Functions	Nonroutine
44.	C	Algebraic Operations	Routine
45.	A	Number Systems	Routine
46.	E	Algebraic Operations	Routine
47.	D	Equations/Inequalities	Nonroutine
48.	C	Algebraic Operations	Routine
49.	B	Equations/Inequalities	Nonroutine
50.	C	Functions	Nonroutine
51.	B	Number Systems	Nonroutine
52.	B	Functions	Routine
53.	−3	Equations/Inequalities	Routine
54.	B	Functions	Routine
55.	B	Algebraic Operations	Nonroutine
56.	D	Functions	Nonroutine
57.	C	Equations/Inequalities	Routine
58.	E	Algebraic Operations	Nonroutine
59.	5	Equations/Inequalities	Nonroutine
60.	E	Functions	Routine

PRACTICE TEST 2

Detailed Explanations of Answers

1. **(C)** Choices (D) and (E) are eliminated because both are irrational.

    (A) $|-5-1| = 6$

    (B) $\left(\dfrac{5}{2}\right)^2 = \dfrac{25}{4} = 6.25$

    (C) $6.\overline{25} = 6.252525...$

2. **(D)**

    I. $\left(\dfrac{-1}{4}\right)^{-2} = \dfrac{(-1)^{-2}}{4^{-2}} = \dfrac{4^2}{(-1)^2} = 16$ or $\left(\dfrac{-1}{4}\right)^{-2} = \dfrac{1}{\left(\dfrac{-1}{4}\right)^2} = \dfrac{1}{\dfrac{1}{16}} = 16$

    II. $(-4)^{-2} = \dfrac{1}{(-4)^2} = \dfrac{1}{16}$

    III. $4^{-1/2} = \dfrac{1}{4^{1/2}} = \dfrac{1}{\sqrt{4}} = \dfrac{1}{2}$

3. **(E)** The two algebraic methods to solve this system are by substitution or elimination.

    Substitution

    $y = 3x - 11$

    $7x + 3(3x - 11) + 1 = 0$

    $7x + 9x - 33 + 1 = 0$

    $16x - 32 = 0$

    $x = 2$

    Eliminating x

    $7x + 3y + 1 = 0$

    $9x - 3y - 33 = 0$

    ──────────────

    $16x - 32 = 0$

    $x = 2$

4. **(D)** The graph is not parabolic, eliminating choices (A) and (B).

    The graph goes through the points $(-3, -1), (-1, 0), (-3, 1), (-3, 2)$.

    Only point $(-1, 0)$ satisfies choices (C).

    Choice (D) satisfies all points.

    Choices (E) is eliminated as none of those points are satisfied.

5. **(B)**

$3x^2 - x - 2 = x^2 - 2x - 1$

$2x^2 + x - 1 = 0$

$(2x - 1)(x + 1) = 0$

$2x - 1 = 0$ $\qquad\qquad x + 1 = 0$

$x = \dfrac{1}{2}$ $\qquad\qquad x = -1$

6. **(D)**

(A) $(4x + 9)^2 = 16x^2 + 72x + 81$

(B) $(-4x - 9)^2 = 16x^2 + 72x + 81$

(C) $|4x + 9|^2$: The value of $4x + 9$ can be either positive or negative, but when it is squared, it is always positive and equal to $16x^2 + 72x + 81$.

(D) $\left(\sqrt{16x^2 + 81}\right)^2 = 16x^2 + 81$

7. **(E)**

$\log_5 4 + \log_5 25 = \log_5 (4 \cdot 25) = \log_5 100$

$2 \log_5 10 = \log_5 10^2 = \log_5 100$

$\log_5 10 - \log_5 \dfrac{1}{10} = \log_5 \dfrac{10}{\frac{1}{10}} = \log_5 (10 \cdot 10) = \log_5 100$

(Note: Although scientific calculators calculate log with base 10 only, when calculating these three values with log base 10 instead of 5, they are still the same.)

8. **(D)** These are all unfamiliar graphs generated by complicated equations. But the only question is: which graph fails the vertical line test and thus is not a function? And only one does. Draw a vertical line in choice (D) just to the right of the origin and it intersects the graph in more than one point.

9. **(A)** This is the associative law: $(ab)(c) = a(bc)$. In this case:

$[(17)(125)](8) = (17)[(125)(8)]$

10. **(D)**

$f(a+1) - f(a-1)$
$= (a+1)^2 - 1 - [(a-1)^2 - 1]$
$= a^2 + 2a + 1 - 1 - (a^2 - 2a + 1 - 1)$
$= a^2 + 2a - a^2 + 2a = 4a$

Alternatively, choosing a value for a can be used here. For example, if $a = 4$, the problem is to find the value of $f(5) - f(3)$.

Then $f(5) = 24$ and $f(3) = 8$, so $f(5) - f(3) = 16$.

The only choice that gives an answer of 16 for $a = 4$ is (D).

11. **(C)** The expansion alternates positive, negative, …, so the third term is positive.

Third term: $_6C_2 (2x)^4 \left(\dfrac{-1}{4y} \right)^2 = \dfrac{6!}{2!4!}(16x^4)\left(\dfrac{1}{16y^2} \right) = \dfrac{(6)(5)x^4}{(2)(1)y^2} = \dfrac{15x^4}{y^2}$

12. **(D)** This example is essentially asking for the intersection of the two sets represented by the inequalities:

$-3 \leq x < 2 : \{-3, -2, -1, 0, 1\}$

$(-1, 4] : \{0, 1, 2, 3, 4\}$

Only two integers, 0 and 1, are in both solutions.

13. **(A)**

$(2^3)^{2x-1} = (2^{-2})^{2-x}$

$2^{6x-3} = 2^{-4+2x}$

$6x - 3 = -4 + 2x$

$4x = -1$

$x = -\dfrac{1}{4}$

This problem can be done by trial-and-error, but it is time-consuming.

14. **(D)**

$|3x| = 9$

$3x = 9$ $\qquad\qquad$ $-3x = 9$

$x = 3$ $\qquad\qquad$ $x = -3$

Check: $\qquad\qquad$ Check:

$|3 \cdot 3| = |9| = 9$ $\qquad$ $|3 \cdot -3| = |-9| = 9$

There are two solutions, $x = 3, -3$.

$|2x - 5| + 3x = 0$

$(2x - 5) + 3x = 0$ $\qquad$ $-(2x - 5) + 3x = 0$

$2x - 5 + 3x = 0$ $\qquad\quad$ $-2x + 5 + 3x = 0$

$5x = 5$ $\qquad\qquad\qquad$ $x + 5 = 0$

$x = 1$ $\qquad\qquad\qquad$ $x = -5$

Check: $\qquad\qquad\qquad$ Check:

$|-3| + 3 = 3 + 3 \ne 0$ $\quad$ $|-15| - 15 = 15 - 15 = 0$

There is one solution, $x = -5$.

So $a = 2$ and $b = 1$, and $a + b = 3$.

15. **(C)**

$$\frac{-6x^{2/3}}{3x^{-3/2}} = -2x^{\left(\frac{2}{3}+\frac{3}{2}\right)} = -2x^{\left(\frac{4}{6}+\frac{9}{6}\right)} = -2x^{13/6}$$

16. **(E)**

$g(f(-2))$: $f(-2) = 0$, $g(0) = 3$

$f(g(-2))$: $g(-2) = -1$, $f(-1) = 1$

$3 - 1 = 2$

17. **(800)**

$R(x) - C(x) = 0$

$\dfrac{x^2}{2} - 400x = 0$

$x^2 - 800x = 0$

$x(x - 800) = 0$

$x = 0, x = 800$

The solution $x = 0$ makes no sense (making and selling no computers). So the company must sell 800 computers to break even.

18. **(D)** At least one real solution means the discriminant $b^2 - 4ac \geq 0$.

   I. $a = 1, b = 20, c = 100$. So $b^2 - 4ac = 400 - 400 = 0$, and there is one real double solution.

   II. $a = 5, b = 2, c = -16$. So $b^2 - 4ac = 4 + 320 = 324 > 0$, and there are two real solutions.

   III. $a = 4, b = -7, c = 4$. So $b^2 - 4ac = 49 - 64 = -15 < 0$, and there are no real solutions.

19. **(E)**

   (A) If $x = -8, y = 0$.

   (B) If $x = -8, y = \dfrac{1}{64+8} = \dfrac{1}{72}$.

   (C) If $x = -8, y = 1$.

   (D) If $x = -8, y = 0$.

   (E) If $x = -8, y = \dfrac{1}{0}$, which does not exist.

20. **(C)**

   I. $\ln e^2 - \ln e^5 = 2 \ln e - 5 \ln e = 2 - 5 = -3$

   II. $2 \ln e^{1/2} = 2\left(\dfrac{1}{2}\right) \ln e = 1$

   III. $\ln 1 = 0$

21. **(E)** Use the slope-intercept form of a line, $y = mx + b$, and the definition of slope $m = \dfrac{\text{rise}}{\text{run}} = \dfrac{2}{3}$ with the given y-intercept, $b = -1$.

   Since the shading is above the line and the line is solid, the inequality is

   $y \geq \dfrac{2}{3}x - 1$.

   Then $3y \geq 2x - 3$

   $0 \geq 2x - 3y - 3$, or $2x - 3y - 3 \leq 0$.

   Trial-and-error also works on this problem by choosing some points and checking them. For example, since (0, 0) is in the shaded region, choices (A), (C), and (D) are eliminated, and since (1, 0) is in the shaded region, choice (B) is eliminated.

22. **(A)**

$(2 - 3i)^2 - (2 - 3i)$
$= (2 - 3i)(2 - 3i) - (2 - 3i)$
$= 4 - 12i + 9i^2 - 2 + 3i$
$= 4 - 12i + 9(-1) - 2 + 3i$
$= 4 - 12i - 9 - 2 + 3i$
$= -7 - 9i$

23. **(E)**

(A) $4^{(3^2)} = 4^9$

(B) $4^{(2^3)} = 4^8$

(C) $16^6 = (4^2)^6 = 4^{12}$

(D) $\left(2^5\right)^2 = \left(4^{1/2}\right)^{10} = 4^5$

(E) $\left(\dfrac{1}{4}\right)^{-4^2} = \left(4^{-1}\right)^{-16} = 4^{16}$

24. **(E)** Since choices (A) and (B) are factorable, they can be eliminated right away. Solving the other choices yields

(C) $x^2 + 1 = 0 \Rightarrow x^2 = -1$ and $x = \pm i$

(D) $x^2 - 2 = 0 \Rightarrow x = \pm\sqrt{2}$

(E) $x^2 - 2x + 2 = 0 \Rightarrow x = \dfrac{2 \pm \sqrt{4 - 4(2)}}{2} = \dfrac{2 \pm \sqrt{-4}}{2} = \dfrac{2 \pm 2i}{2} = 1 \pm i$

Note that this problem can also be done by trial-and-error by plugging $1 - i$ into each equation, but this is time-consuming.

25. **(C)**

$y = \dfrac{3x - 5}{2x + 1}$ so the inverse is $x = \dfrac{3y - 5}{2y + 1}$. Solve for y:

$2xy + x = 3y - 5$

$x + 5 = 3y - 2xy$

$x + 5 = y(3 - 2x)$

$y = \dfrac{x+5}{3-2x}$

26. **(B)** For short sums such as this one with only four terms, write out each term and add.

$$\sum_{n=0}^{3}(-1)^n \dfrac{1}{2^n} = (-1)^0 \dfrac{1}{2^0} + (-1)^1 \dfrac{1}{2^1} + (-1)^2 \dfrac{1}{2^2} + (-1)^3 \dfrac{1}{2^3}$$

$$= 1 - \dfrac{1}{2} + \dfrac{1}{4} - \dfrac{1}{8}$$

$$= 1\left(\dfrac{8}{8}\right) - \dfrac{1}{2}\left(\dfrac{4}{4}\right) + \dfrac{1}{4}\left(\dfrac{2}{2}\right) - \dfrac{1}{8}$$

$$= \dfrac{8 - 4 + 2 - 1}{8}$$

$$= \dfrac{5}{8}$$

27. **(C)** To find the remainder, evaluate $x^5 - 10x + 2$ at $x = -2$.

$(-2)^5 - 10(-2) + 2 = -32 + 20 + 2 = -10$

28. **(A)** Cubic equations can sometimes be factored by looking at them as two factorable parts:

$-x^3 + 4x^2 + 4x - 16$

$= -[(x^3 - 4x^2) + (-4x + 16)]$

$= -[x^2(x - 4) - 4(x - 4)]$

$= -(x - 4)(x^2 - 4)$

$= -(x - 4)(x + 2)(x - 2)$

The problem also can be done by trial-and-error.

Choices (B) and (E) can be eliminated as they are 4th-degree expressions.

For (A): $-(x - 4)(x + 2)(x - 2)$

$= -(x - 4)(x^2 - 4)$

$= -(x^3 - 4x^2 - 4x + 16)$

$= -x^3 + 4x^2 + 4x - 16$

29. **(B)**

$$\left(\frac{3+i}{i}\right)\left(\frac{i}{i}\right) = \frac{3i+i^2}{i^2} = \frac{3i-1}{-1} = 1-3i$$

The 1 is the real part (horizontal) and the -3 is the imaginary part (vertical).

30. **(C)** There are three ways to do this problem.

Directly: Find the slope and use it to find the general equation:

$$m = \frac{1-(-1)}{3-2} = 2$$

$y - 1 = 2(x - 3)$

$y - 1 = 2x - 6$

$2x - y = 5$

$M = 2, N = -1$

Directly: Solve the system of equations for M and N:

$(3, 1) : 3M + N = 5$

$(2, -1) : 2M - N = 5$

$5M = 10$ and $M = 2$

$3(2) + N = 5$, so $N = -1$

$M = 2, N = -1$

Trial-and-error: Plug M and N into the equation and check whether both points are solutions. This method is fairly time-consuming, but it works.

31. **(B)** Rewrite as $z - y = x$ and $y - z = -x$, and plug these into $|z - y| + |y - z|$:

$$|z - y| + |y - z| = |x| + |-x| = |x| + |x| = 2|x|$$

Note that we cannot say $2|x| = 2x$. For instance, when $x = 3$, it is true that $|x| + |-x| = |3| + |-3| = 6 = 2(3)$. But when $x = -3$, $|x| + |-x| = |-3| + |3| = 6 \neq 2(-3)$.

Also, be careful about using trial-and-error with absolute value problems.

Using $x = 2, y = 3, z = 5$ might lead to choice (A), but $x = -2, y = 3, z = 1$ would show that choice (A) isn't right.

PRACTICE TEST 2: DETAILED EXPLANATIONS OF ANSWERS | 323

32. **(728)** The sequence is $\{16, 16, 18, 18, 20, 20, ...\}$

This is two arithmetic sequences of 16, 18, 20, ... to give two sets of seats with 13 rows each.

Use $a_n = a_1 + (n-1)d$ with $a_1 = 16, d = 2, n = 13$

$a_{13} = 16 + 12(2) = 40$

Then the sum for one of the arithmetic sequences is given by $S_n = \dfrac{n}{2}(a_1 + a_n)$, or $S_{13} = \dfrac{13}{2}(16 + 40) = 364$.

So there are $2(364) = 728$ seats.

Using a calculator to find $16 + 16 + 18 + 18 + ... + 40 + 40$ would be time-consuming, and you would still have to determine how many seats are in the last row.

33. **(D)**

$\sqrt{2x+14} = x+3$

$\left(\sqrt{2x+14}\right)^2 = (x+3)^2$

$2x + 14 = x^2 + 6x + 9$

$0 = x^2 + 6x + 9 - 2x - 14$

$0 = x^2 + 4x - 5$

$0 = (x+5)(x-1)$

$x + 5 = 0 \qquad x - 1 = 0$

$x = -5 \qquad x = 1$

Checking for extraneous solutions by using the original equation is necessary for radical equations as well as absolute value equations.

Check: $x = -5 \Rightarrow \sqrt{-10+14} + 5 = 7 \neq 3$

Check: $x = 1 \Rightarrow \sqrt{2+14} - 1 = 4 - 1 = 3$

So there is only one solution: $x = 1$.

34. **(B)**

   I. $144^{-1/2} = \dfrac{1}{\sqrt{144}} = \dfrac{1}{12}$ (square roots are always positive).

   II. $(64)^{2/3} = \left(\sqrt[3]{64}\right)^2 = 4^2 = 16$

   III. $(-32)^{-3/5} = \dfrac{1}{(-32)^{3/5}} = \dfrac{1}{\left(\sqrt[5]{-32}\right)^3} = \dfrac{1}{(-2)^3} = \dfrac{-1}{8}$

35. **(69)**

   This is an arithmetic sequence with $a_1 = 902$, $d = -17$, and $n = 50$.

   $a_n = a_1 + (n-1)d$

   $a_{50} = 902 + 49(-17) = 902 - 833 = 69$

36. **(D)**

   I. $\log_9 1 = x$, so $9^x = 1 \Rightarrow x = 0$ since the log of 1 for any base is 0.

   II. $\log_9 \dfrac{1}{3} = x$, so $9^x = \dfrac{1}{3} \Rightarrow 3^{2x} = 3^{-1}$ and $x = \dfrac{-1}{2}$.

   III. $\log_{27} 9 = x$, so $27^x = 9 \Rightarrow 3^{3x} = 3^2$ and $x = \dfrac{2}{3}$.

37. **(D)**   Step-by-step, the solution is

   $f(2) = 2 - 2^2 = 2 - 4 = -2$

   $f(-2) = 2(-2) + 3 = -4 + 3 = -1$

   $f(-1) = -(-1) - 1 = 0$

38. **(E)**   Choice (A) is arithmetic with a common difference $d = 6$.

   Although choices (B), (C), and (D) have patterns, they are not geometric because there is no common ratio.

   Choice (E) is geometric with $r = -\dfrac{1}{4}$.

PRACTICE TEST 2: DETAILED EXPLANATIONS OF ANSWERS | 325

39. **(E)**

$$4 \leq 3 - \frac{1}{2}x < \frac{7}{4}$$
$$16 \leq 12 - 2x < 7$$
$$4 \leq -2x < -5$$
$$-4 \geq 2x > 5$$
$$-2 \geq x > \frac{5}{2}$$
$$(-\infty, -2] \cup \left(\frac{5}{2}, \infty\right)$$

40. **(A)**

$$\text{LCD} = (a+b)(a-b)$$
$$\frac{5}{a+b}\left(\frac{a-b}{a-b}\right) - \frac{4}{a-b}\left(\frac{a+b}{a+b}\right)$$
$$= \frac{5a - 5b - 4a - 4b}{(a+b)(a-b)}$$
$$= \frac{a - 9b}{a^2 - b^2}$$

41. **(B)**

$x^2 = 3x + 18$

$x^2 - 3x - 18 = 0$

$(x - 6)(x + 3) = 0$

$x = 6, x = -3$

However, a base cannot be negative, so $x = 6$ is the only solution.

42. **(E)**

I. is not true; $\sqrt{3} \cdot \sqrt{5} = \sqrt{15}$, which is irrational.

II. is true; $\sqrt{\frac{1}{2}} \cdot \sqrt{\frac{1}{2}} = \frac{1}{2}$.

III. is true; $\sqrt{2} \cdot \sqrt{2} = 2$.

43. **(C)** For this curve to have three roots, it would have to cross or touch the *x*-axis in three points. This would happen if the graph was moved down more than 1 unit, and the only transformation that would do that is (C).

This problem can also be solved by the process of elimination. Choices (A) and (D) aren't correct because they would only move the graph to the right or the left. Choice (E) is eliminated because it would "flip" the curve but still cross the *x*-axis only once. Choice (B) is eliminated because it moves (1, 1) down to the *x*-axis, but then it crosses or touches the axis in only two points. The correct answer is choice (C).

44. **(C)**

LCD = 36

$$36\left(\frac{5x+2}{6} - \frac{3x-1}{9} = \frac{2x+3}{12}\right)$$

$6(5x + 2) - 4(3x - 1) = 3(2x + 3)$

$30x + 12 - 12x + 4 = 6x + 9$

$18x + 16 = 6x + 9$

$12x = -7$

$x = \dfrac{-7}{12}$

Doing this problem by trial-and-error is too time-consuming.

45. **(A)**

I. $\begin{vmatrix} -9 & -9 \\ -9 & -9 \end{vmatrix} = -9(-9) - (-9)(-9) = 81 - 81 = 0$

II. $\begin{vmatrix} 9 & -9 \\ -9 & -9 \end{vmatrix} = 9(-9) - (-9)(-9) = -81 - 81 = -162$

III. $\begin{vmatrix} \dfrac{-1}{9} & \dfrac{2}{3} \\ \dfrac{3}{2} & 9 \end{vmatrix} = \dfrac{-1}{9}(9) - \dfrac{3}{2}\left(\dfrac{2}{3}\right) = -1 - 1 = -2$

46. **(E)**

$$\frac{-1\cancel{(x-2)}}{x\cancel{(x+3)}} \cdot \frac{\cancel{(x+3)}\cancel{(x+1)}}{\cancel{(x-2)}\cancel{(x+1)}} = \frac{-1}{x}$$

All the factors in parentheses cancel.

47. **(D)** Since all terms in both equations have squares in them, it doesn't matter whether x and y are positive or negative. So since (2, 3) is a solution, so are (2, –3), (–2, 3), and (–2, –3), for a total of three more. Choice (E) gives the *total* number of solutions, but the problem asks how many *additional* solutions.

48. **(C)**

LCD = 10

$$10\left(\frac{2x}{5} - \frac{5x}{2} \geq 1 - x\right)$$

$4x - 25x \geq 10 - 10x$

$-11x \geq 10$

$x \leq \frac{-10}{11}$

49. **(B)** The graphs share the same y-intercept (0, 4), and the shading in both is above the line. One line has a positive slope and the other has a negative slope, so their common intersections must be in quadrants I and II only (see figure below).

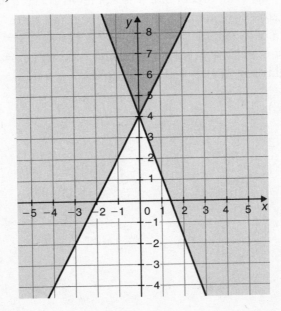

50. **(C)** The inverse function switches the x- and y-values, so it will pass through these points.

x	4	0	−4
$f^{-1}(x)$	−4	0	4

Only (C) passes through these points.

51. **(B)**

$$\left(\frac{1+i}{4+2i}\right)\left(\frac{4-2i}{4-2i}\right)$$

$$=\frac{4-2i+4i-2i^2}{16-4i^2}$$

$$=\frac{4+2i-2(-1)}{16-4(-1)}$$

$$=\frac{4+2i+2}{16+4}$$

$$=\frac{6+2i}{20}$$

$$=\frac{6}{20}+\frac{2i}{20}$$

$$=\frac{3}{10}+\frac{1}{10}i, \text{ so } a=\frac{3}{10}.$$

52. **(B)** Since $f(x)$ is a polynomial, it is a smooth curve. The roots are points where $f(x)$ crosses or touches the x-axis, or ($f(x) = 0$). The function $f(x)$ changes sign between $x = -4$ and $x = -2$, between $x = 4$ and $x = 6$, and between $x = 6$ and $x = 8$, so it crosses the x-axis ($f(x) = 0$) between those points. In addition, $f(x) = 0$ at $x = 2$, so there is a minimum of four roots.

53. (−3)

Divide the known equations to get

$$\frac{xy^6}{xy^3} = \frac{-192}{24}$$

$y^3 = -8$

$y = -2$

Substitute $y = -2$ into either original equation:

$x(-2)^3 = 24$ or $x(-2)^6 = -192$

$-8x = 24$ $64x = -192$

$x = -3$ $64x = -192$

                               $x = -3$

54. **(B)**

$a = f(g(-2)) = f(4) = 14$ $b = g(f(-2)) = g(8) = -6$

$c = f(g(2)) = f(0) = 2$ $d = g(f(2)) = g(4) = -2$

$c + d = 0$ but $a + b \neq 0$.

55. **(B)**

$x - x + x(x - y) - y(x - y)$

$= x - x + x^2 - xy - xy + y^2$

$= x^2 - 2xy + y^2$

$= (x - y)^2$

56. **(D)**

The slope of line k is $m = \dfrac{10-1}{-12-6} = \dfrac{9}{-18} = -\dfrac{1}{2}$.

So the slope of line l is the negative reciprocal, or $m = 2$.

The equation of line l is thus $y = 2x + 6$.

The x-intercept is at $y = 0 = 2x + 6$, so $x = -3$.

57. **(C)**

$(6x - 3) < 15 \qquad -(6x - 3) < 15$

$6x - 3 < 15 \qquad -6x + 3 < 15$

$6x < 18 \qquad\qquad -6x < 12$

$x < 3 \qquad\qquad\quad x > -2$

So $-2 < x < 3$ or $(-2, 3)$

Check: $x = 0$: $|6(0) - 3| - 15 = 3 - 15 = -12 < 0$

58. **(E)**

$\left(\dfrac{1}{1-\sqrt{2}}\right)^2 = \dfrac{1}{1 - 2\sqrt{2} + 2}$

$= \left(\dfrac{1}{3 - 2\sqrt{2}}\right)\left(\dfrac{3 + 2\sqrt{2}}{3 + 2\sqrt{2}}\right)$

$= \dfrac{3 + 2\sqrt{2}}{9 - 4(2)}$

$= 3 + 2\sqrt{2}$

59. **(5)**

The projectile strikes the ground at $h(t) = 0$.

$h(t) = -16t^2 + 48t + 160 = 0$

$-16(t^2 - 3t - 10) = 0$

$-16(t - 5)(t + 2) = 0$

$t = 5, t = -2$, but time cannot be negative here, so the only answer is $t = 5$.

60. **(E)** The first transformation is reflected across the x-axis, which is $-g(x)$. Then the graph is translated 1 unit to the left, changing it to $y = -g(x + 1)$.

PRACTICE TEST 1

Answer Sheet

1. Ⓐ Ⓑ Ⓒ Ⓓ Ⓔ
2. Ⓐ Ⓑ Ⓒ Ⓓ Ⓔ
3. Ⓐ Ⓑ Ⓒ Ⓓ Ⓔ
4. Ⓐ Ⓑ Ⓒ Ⓓ Ⓔ
5. Ⓐ Ⓑ Ⓒ Ⓓ Ⓔ
6. [_____]
7. Ⓐ Ⓑ Ⓒ Ⓓ Ⓔ
8. Ⓐ Ⓑ Ⓒ Ⓓ Ⓔ
9. Ⓐ Ⓑ Ⓒ Ⓓ Ⓔ
10. Ⓐ Ⓑ Ⓒ Ⓓ Ⓔ
11. Ⓐ Ⓑ Ⓒ Ⓓ Ⓔ
12. [_____]
13. Ⓐ Ⓑ Ⓒ Ⓓ Ⓔ
14. Ⓐ Ⓑ Ⓒ Ⓓ Ⓔ
15. Ⓐ Ⓑ Ⓒ Ⓓ Ⓔ
16. Ⓐ Ⓑ Ⓒ Ⓓ Ⓔ
17. Ⓐ Ⓑ Ⓒ Ⓓ Ⓔ
18. Ⓐ Ⓑ Ⓒ Ⓓ Ⓔ
19. Ⓐ Ⓑ Ⓒ Ⓓ Ⓔ
20. Ⓐ Ⓑ Ⓒ Ⓓ Ⓔ
21. [_____]
22. Ⓐ Ⓑ Ⓒ Ⓓ Ⓔ
23. Ⓐ Ⓑ Ⓒ Ⓓ Ⓔ
24. Ⓐ Ⓑ Ⓒ Ⓓ Ⓔ
25. Ⓐ Ⓑ Ⓒ Ⓓ Ⓔ
26. Ⓐ Ⓑ Ⓒ Ⓓ Ⓔ
27. Ⓐ Ⓑ Ⓒ Ⓓ Ⓔ
28. Ⓐ Ⓑ Ⓒ Ⓓ Ⓔ
29. Ⓐ Ⓑ Ⓒ Ⓓ Ⓔ
30. Ⓐ Ⓑ Ⓒ Ⓓ Ⓔ
31. Ⓐ Ⓑ Ⓒ Ⓓ Ⓔ
32. Ⓐ Ⓑ Ⓒ Ⓓ Ⓔ
33. Ⓐ Ⓑ Ⓒ Ⓓ Ⓔ
34. Ⓐ Ⓑ Ⓒ Ⓓ Ⓔ
35. Ⓐ Ⓑ Ⓒ Ⓓ Ⓔ
36. Ⓐ Ⓑ Ⓒ Ⓓ Ⓔ
37. Ⓐ Ⓑ Ⓒ Ⓓ Ⓔ
38. Ⓐ Ⓑ Ⓒ Ⓓ Ⓔ
39. Ⓐ Ⓑ Ⓒ Ⓓ Ⓔ
40. [_____]
41. Ⓐ Ⓑ Ⓒ Ⓓ Ⓔ
42. Ⓐ Ⓑ Ⓒ Ⓓ Ⓔ
43. Ⓐ Ⓑ Ⓒ Ⓓ Ⓔ
44. Ⓐ Ⓑ Ⓒ Ⓓ Ⓔ
45. Ⓐ Ⓑ Ⓒ Ⓓ Ⓔ
46. Ⓐ Ⓑ Ⓒ Ⓓ Ⓔ
47. Ⓐ Ⓑ Ⓒ Ⓓ Ⓔ
48. Ⓐ Ⓑ Ⓒ Ⓓ Ⓔ
49. Ⓐ Ⓑ Ⓒ Ⓓ Ⓔ
50. Ⓐ Ⓑ Ⓒ Ⓓ Ⓔ
51. Ⓐ Ⓑ Ⓒ Ⓓ Ⓔ
52. Ⓐ Ⓑ Ⓒ Ⓓ Ⓔ
53. Ⓐ Ⓑ Ⓒ Ⓓ Ⓔ
54. Ⓐ Ⓑ Ⓒ Ⓓ Ⓔ
55. Ⓐ Ⓑ Ⓒ Ⓓ Ⓔ
56. Ⓐ Ⓑ Ⓒ Ⓓ Ⓔ
57. Ⓐ Ⓑ Ⓒ Ⓓ Ⓔ
58. Ⓐ Ⓑ Ⓒ Ⓓ Ⓔ
59. Ⓐ Ⓑ Ⓒ Ⓓ Ⓔ
60. Ⓐ Ⓑ Ⓒ Ⓓ Ⓔ

PRACTICE TEST 2

Answer Sheet

1. Ⓐ Ⓑ Ⓒ Ⓓ Ⓔ
2. Ⓐ Ⓑ Ⓒ Ⓓ Ⓔ
3. Ⓐ Ⓑ Ⓒ Ⓓ Ⓔ
4. Ⓐ Ⓑ Ⓒ Ⓓ Ⓔ
5. Ⓐ Ⓑ Ⓒ Ⓓ Ⓔ
6. Ⓐ Ⓑ Ⓒ Ⓓ Ⓔ
7. Ⓐ Ⓑ Ⓒ Ⓓ Ⓔ
8. Ⓐ Ⓑ Ⓒ Ⓓ Ⓔ
9. Ⓐ Ⓑ Ⓒ Ⓓ Ⓔ
10. Ⓐ Ⓑ Ⓒ Ⓓ Ⓔ
11. Ⓐ Ⓑ Ⓒ Ⓓ Ⓔ
12. Ⓐ Ⓑ Ⓒ Ⓓ Ⓔ
13. Ⓐ Ⓑ Ⓒ Ⓓ Ⓔ
14. Ⓐ Ⓑ Ⓒ Ⓓ Ⓔ
15. Ⓐ Ⓑ Ⓒ Ⓓ Ⓔ
16. Ⓐ Ⓑ Ⓒ Ⓓ Ⓔ
17. ▭
18. Ⓐ Ⓑ Ⓒ Ⓓ Ⓔ
19. Ⓐ Ⓑ Ⓒ Ⓓ Ⓔ
20. Ⓐ Ⓑ Ⓒ Ⓓ Ⓔ
21. Ⓐ Ⓑ Ⓒ Ⓓ Ⓔ
22. Ⓐ Ⓑ Ⓒ Ⓓ Ⓔ
23. Ⓐ Ⓑ Ⓒ Ⓓ Ⓔ
24. Ⓐ Ⓑ Ⓒ Ⓓ Ⓔ
25. Ⓐ Ⓑ Ⓒ Ⓓ Ⓔ
26. Ⓐ Ⓑ Ⓒ Ⓓ Ⓔ
27. Ⓐ Ⓑ Ⓒ Ⓓ Ⓔ
28. Ⓐ Ⓑ Ⓒ Ⓓ Ⓔ
29. Ⓐ Ⓑ Ⓒ Ⓓ Ⓔ
30. Ⓐ Ⓑ Ⓒ Ⓓ Ⓔ
31. Ⓐ Ⓑ Ⓒ Ⓓ Ⓔ
32. ▭
33. Ⓐ Ⓑ Ⓒ Ⓓ Ⓔ
34. Ⓐ Ⓑ Ⓒ Ⓓ Ⓔ
35. ▭
36. Ⓐ Ⓑ Ⓒ Ⓓ Ⓔ
37. Ⓐ Ⓑ Ⓒ Ⓓ Ⓔ
38. Ⓐ Ⓑ Ⓒ Ⓓ Ⓔ
39. Ⓐ Ⓑ Ⓒ Ⓓ Ⓔ
40. Ⓐ Ⓑ Ⓒ Ⓓ Ⓔ
41. Ⓐ Ⓑ Ⓒ Ⓓ Ⓔ
42. Ⓐ Ⓑ Ⓒ Ⓓ Ⓔ
43. Ⓐ Ⓑ Ⓒ Ⓓ Ⓔ
44. Ⓐ Ⓑ Ⓒ Ⓓ Ⓔ
45. Ⓐ Ⓑ Ⓒ Ⓓ Ⓔ
46. Ⓐ Ⓑ Ⓒ Ⓓ Ⓔ
47. Ⓐ Ⓑ Ⓒ Ⓓ Ⓔ
48. Ⓐ Ⓑ Ⓒ Ⓓ Ⓔ
49. Ⓐ Ⓑ Ⓒ Ⓓ Ⓔ
50. Ⓐ Ⓑ Ⓒ Ⓓ Ⓔ
51. Ⓐ Ⓑ Ⓒ Ⓓ Ⓔ
52. Ⓐ Ⓑ Ⓒ Ⓓ Ⓔ
53. ▭
54. Ⓐ Ⓑ Ⓒ Ⓓ Ⓔ
55. Ⓐ Ⓑ Ⓒ Ⓓ Ⓔ
56. Ⓐ Ⓑ Ⓒ Ⓓ Ⓔ
57. Ⓐ Ⓑ Ⓒ Ⓓ Ⓔ
58. Ⓐ Ⓑ Ⓒ Ⓓ Ⓔ
59. ▭
60. Ⓐ Ⓑ Ⓒ Ⓓ Ⓔ

Glossary

Absolute value: The distance of a number from zero on a number line; never negative.

Absolute value curve: The graph of $f(x) = |x|$ which is a V-shaped curve.

Additive identity: The number zero; the sum of zero and any number is that number.

Additive inverse: The negative of a given number. The additive inverse of a is $-a$.

Arithmetic means: For any set of data, the sum of all numbers divided by the number of numbers.

Arithmetic series: The sum of the terms of an arithmetic sequence.

Arithmetic sequence: A sequence in which the difference (d) between two consecutive terms is the same.

Associative property: The sum (or product) of any three real numbers is the same regardless of the way they are grouped. $(a + b) + c = a + (b + c)$ and $(ab)c = a(bc)$.

Base: A number used as a repeated factor.

Binomial: A polynomial that has two terms.

Binomial expansion: Writing an expression in the form $(x + y)^n$ as a series of terms, where n is an integer.

Closure property: The sum (or product) of any two real numbers is always a real number.

Coefficient: The number that is multiplied by the variables of a term.

Combination: The number of ways n items can be placed in groups of r and written as $_nC_r$.

Common denominator: The quantity divisible by every denominator in the equation or expression.

Common difference: The constant difference (d) between terms in an arithmetic sequence.

Common logarithm: A logarithm with base 10.

Common ratio: The constant ratio r between terms in a geometric sequence.

Commutative property: The sum (or product) of any two real numbers is the same even if their positions are changed. $a + b = b + a$ and $ab = ba$.

Complement of a set: The complement of set A, denoted as A', is all elements of the universal set not contained in A.

Complex conjugates: $a + bi$ and $a - bi$ are complex conjugates. Multiplying an expression by its complex conjugate will eliminate the imaginary part.

Complex fraction: A fraction in which either the numerator or the denominator, or both, are rational expressions (a fraction within a fraction).

Complex number: A number of the form $a + bi$ where a and b are real numbers.

Composition of a function: Application of a function to the results of another function. Typically written as $f(g(x))$.

Conjugate: $a + \sqrt{b}$ and $a - \sqrt{b}$ where a is a real number and b is a positive real number are conjugates. When conjugates are multiplied, the square root term disappears.

Consistent system of linear equations: A system of linear equations that has one solution.

Constant: A quantity that does not change.

Constant function: A linear function whose slope $m = 0$ and is in the form $y = b$ where b is the y-intercept.

Convergent series: An infinite geometric series whose sum is finite.

Cube: To raise an expression to the 3rd power.

Cube root: A number whose cube is a given number.

Curve: The shape of the graph of a relation. Not necessarily curved.

Decay curve: A curve whose equation is in the form of $y = ab^x$, $0 < b < 1$.

Degree of a polynomial: The highest degree of its terms.

Dependent system of linear equations: A system of linear equations with an infinite number of solutions. This happens when the two lines are the same.

Determinant: A value associated with a square matrix. The determinant of $\begin{vmatrix} a & b \\ c & d \end{vmatrix}$ is $ad - bc$.

Dimension of a matrix: Expressed as the number of rows of the matrix by the number of columns of the matrix. A matrix with 3 rows and 2 columns would have dimension 3×2.

Discriminant: The value of $b^2 - 4ac$ in a quadratic equation $a^2 + bx + c = 0$. The value of the discriminant is an indicator of how many solutions the quadratic equation has.

Disjoint: Two sets are disjoint if they have no elements in common.

Distributive property: The distributive property says that for all real numbers, $a(b \pm c) = ab \pm ac$.

Domain: The set of all the values of x in a relation. Domains are typically all real numbers with the exception of values of x creating a zero in the denominator, a square root of a negative number, or the logarithm of a negative number.

e: A special irrational constant used as the base of the natural logarithm function. $e \approx 2.71828\ldots$

Element: Each member or value in a matrix or set.

Elimination method: A method to solve systems of simultaneous equations by multiplying one or both equations by a constant to eliminate a variable.

Equation: Two algebraic expressions separated by an equal sign, showing that the two sides have equal values.

Even: The set of integers $\{\ldots, -6, -4, -2, 0, 2, 4, 6, \ldots\}$ that are divisible by 2. Zero is considered to be even.

Even function: A function that is symmetric to the y-axis. In even functions, for all values a, $f(-a) = f(a)$.

Exponent: The number of times a factor is used.

Exponential function: A function in the form of ab^x where a is a real number not equal to zero and b is a positive real number.

Expression: A collection of one or more terms.

Extraneous solution: A solution that, when plugged back into the original equation, does not give a true statement. Extraneous solutions can occur when squaring both sides of an equation.

Factor: One or more expressions that when multiplied together, produce a given result.

Factor theorem: If a is a root of the equation, then $(x - a)$ is a factor of $f(x) = 0$.

Factorial: $n! = n(n - 1)(n - 2)\ldots(3)(2)(1)$ where n is a whole number.

Factoring: The process of rewriting an expression as a product.

Finite set: A set with a countable number of members.

FOIL: An acronym for multiplying two binomials $(x + a)(x + b)$. FOIL stands for Firsts-Outers-Inners-Lasts.

Fractional exponents: A fractional exponent of an expression creates a root. An exponent in the form of $a^{m/n} = \sqrt[n]{a^m} = \left(\sqrt[n]{a}\right)^m$.

Function: A relation such that for all x in the domain, there is one and only one y in the range. In functions, the y can repeat but the x cannot.

Function notation: A function written in the form $y = f(x)$ gives a rule for finding the value of y at a given value of x.

General term of a sequence: A formula for the nth term of a sequence. This term can be found by plugging in any whole number n into this general term.

Geometric sequence: A sequence in which the ratio r between two consecutive terms is the same.

Geometric series: The sum of n terms of a geometric sequence.

Growth curve: A curve whose equation is in the form of $y = ab^x, b > 1$.

Horizontal line: An equation in the form of $y = b$ creates a horizontal line.

Horizontal line test: A method to determine when, given a function $f(x)$ in graphical form, its inverse $f^{-1}(x)$ is also a function. If any horizontal line does not intersect the graph of $f(x)$ in more than one point, $f^{-1}(x)$ is also a function.

Imaginary number: Any number in the form of bi where b is a real number, $b \neq 0$ and $i = \sqrt{-1}$.

Imaginary root: A root of a quadratic equation containing an imaginary number.

Improper fraction: A fraction in the form $\frac{b}{a}, b \geq a$.

Inconsistent solution: A system of linear equations with no solutions. This happens when the two lines are parallel.

Index: The root of a radical. Square roots have index 2, cube roots have index 3.

Inequality: A mathematical statement that two quantities are not equal. Inequalities use the signs $<, \leq, >,$ and $\geq$.

Infinite set: A set with an infinite number of elements.

Integers: The set of natural numbers, the negative natural numbers, and zero. $\{\ldots, -3, -2, -1, 0, 1, 2, 3, \ldots\}$.

Intercept form of a linear function: A linear equation in the form of $\frac{x}{a} + \frac{y}{b} = 1$ where a is the x-intercept and b is the y-intercept where $a \neq 0, b \neq 0$.

Intersection of graphs: The point or points two graphs have in common.

Intersection of sets: Given two sets A and B, the set of elements that belong to both A and B. Written as $A \cap B$.

Interval notation: A method of writing inequalities using intervals where brackets [ ] include end values and parentheses ( ) do not include end values.

Inverse of a function: The inverse of a function in the form of points (x, y) is the set of points (y, x). Inverses can be found by switching the values of x and y when the function is given using function notation.

Irrational number: A number that cannot be expressed as $\frac{a}{b}$, the ratio of two integers. Non-terminating, non-repeating decimals are irrational as well as numbers like $\pi, e,$ or $\sqrt{2}$.

Like terms: Like terms are terms that are added having the same variables and exponents.

Linear equation: An equation in the form of $ax + by = c$, where a, b, c are constants and a and b are not simultaneously zero.

Linear function: A function in the form of $y = mx + b$ which graphs a straight line.

Linear inequality: An equation in the form of $ax + by > c$ or $ax + by < c$ that will contain all points above or below the line.

Logarithm: A logarithm solves for the exponent of an exponential equation. $\log_b y = x$ is equivalent to $b^x = y$.

Logarithmic equation: An equation involving logarithms. Usually, the equation is solved by rewriting it exponentially.

Long division: The division of a polynomial by a polynomial.

Lowest common denominator (LCD): The smallest quantity divisible by every denominator in the equation or expression.

Matrix: A rectangular arrangement of numbers given in rows and columns.

Monomial: An algebraic expression consisting of a coefficient, variables and powers.

Multiplicative identity: The number 1; multiplying 1 by any number gives that number.

Multiplicative inverse: The reciprocal of a number a which is $\frac{1}{a}$. Not to be confused with the inverse of a function.

Multiplicity: When an expression is factored, the multiplicity of a factor is the number of times that factor is used. Since $18 = 2(3)(3)$, the multiplicity of 2 is one and the multiplicity of 3 is two.

Natural number: A number from the set $\{1, 2, 3, ...\}$.

Negative integral exponent: An expression with a negative exponent creates a fraction. The definition of a^{-n} is $\frac{1}{a^n}$.

Negative rational exponent: An expression with a negative rational exponent is the reciprocal of the expression with the positive rational expression $a^{-m/n} = \frac{1}{a^{m/n}}$.

***n*th term of a sequence:** A formula for the general term of the sequence in terms of n.

Null set: The set containing no elements denoted as $\emptyset$, also called the empty set.

Odd: The set of integers $\{...,-7, -5, -3, -1, 1, 3, 5, 7, ...\}$ which are not divisible by 2. Zero is considered to be even.

Odd function: A function that is symmetric to the origin. In odd functions, for all values $a, f(-a) = -f(a)$.

Operations: A process applied to algebraic expressions that may change them. Addition, subtraction, multiplication, division, squaring are some examples of operations.

Order of operations: When multiple operations are given in an expression, the order of operations is a rule that generalizes which to do first. Parentheses are resolved first, and then exponents. Multiplications and divisions are performed left to right and then addition and subtraction, left to right.

Ordered pair: A pair of numbers in the form of (x, y) used to locate a number on the coordinate plane. The first number x tells how far to move horizontally and the second number y tells how far to move vertically.

Origin: The point $(0,0)$ in the coordinate plane.

Origin symmetry: A function which is a mirror image across the origin. In origin symmetry, points that are in quadrant I are reflected to quadrant III and vice versa. Also, points that are in quadrant II are reflected to quadrant IV and vice versa.

Parabola: The u-shaped curve that is graphed by a quadratic equation $y = ax^2 + bx + c$ where a, b, c are real numbers and $a \neq 0$.

Parallel lines: Lines that have no intersection point. Parallel lines have the same slope and different y-intercepts.

Pascal's triangle: A triangular array of the coefficients used in binomial expansion.

Perfect cube: A number that represents the cube of natural numbers. Perfect cubes are 1, 8, 27, 64, 125, ...

Perfect square: A number that represents the square of natural numbers. Perfect squares are 1, 4, 9, 16, 25, 36,...

Perpendicular lines: Two lines that form a right angle at their intersection. When two lines are perpendicular, their slopes are negative reciprocals.

Piecewise function: A function defined with two or more rules, depending on the value of x.

Point-slope form: An equation of a line in the form of $y - y_1 = m(x - x_1)$ where m is the slope of the line and (x_1, y_1) is a point on the line.

Polynomial: An algebraic expression with one or more terms.

Power: Same as exponent.

Prime: Two numbers or algebraic expressions are prime if they cannot be factored other than by one and itself.

Quadrant: Sections of the coordinate plane formed by the intersection of the x- and y-axes. There are four quadrants.

Quadratic equation: A second-degree equation in x in the form of $y = ax^2 + bx + c$ where a, b, c are real numbers and $a \neq 0$.

Quadratic formula: A formula to solve a quadratic equation in the form of $y = ax^2 + bx + c$. The quadratic formula states that $x = \dfrac{-b \pm \sqrt{b^2 - 4ac}}{2a}$.

Quadratic inequality: An inequality involving a quadratic. The solution will be the area either inside or outside the graph of the quadratic equation parabola.

Quotient: An expression described as $\dfrac{f(x)}{g(x)}, g(x) \neq 0$.

Radical: The designated symbol $\sqrt[n]{}$ for the nth root of a mathematical expression.

Radical Equation: An equation that has one or more variables under a radical.

Radicand: The expression inside a radical $\sqrt{}$.

Range: The set of all y values in a relation.

Rational exponent: An exponent in the form of $a^{m/n} = \sqrt[n]{a^m} = \left(\sqrt[n]{a}\right)^m$.

Rational expression: An algebraic expression that can be written as the quotient of two polynomials.

Rational number: A number that can be expressed as $\dfrac{a}{b}$, the ratio of two integers, $b \neq 0$. All fractions and terminating or repeating decimals are rational.

Rationalize: To eliminate radicals from the denominator. Multiplying numerator and denominator by the denominator's conjugate will rationalize a quotient.

Real number: Numbers that are either rational or irrational.

Real roots: Values of x that satisfy the equation $f(x) = 0$. Also known as solutions or zeros of the function.

Reciprocal: The reciprocal of a number a is $\dfrac{a}{b}$.

Reciprocal function: The reciprocal of a function $f(x)$ is $\dfrac{1}{f(x)}$. Not to be confused with the inverse of a function.

Relation: A set of ordered pairs (x, y).

Remainder theorem: If a is any constant and if the polynomial $f(x)$ is divided by $(x - a)$, the remainder is $f(a)$.

Root of an equation: A solution to an equation $f(x) = 0$.

Scientific notation: A number written in the form of a product of a real number $a (1 \leq a < 10)$ and an integral power of 10. Used for very large or very small numbers.

Sequence: A set of numbers written as the 1st term, 2nd term, 3rd term, ... Usually there is a logical pattern to the sequence with a formula for the general or n^{th} term.

Series: The sum of a sequence of n terms.

Set: A group of items or elements.

Slope of a line: The measure of the steepness of a line in the form of $m = \dfrac{\text{rise}}{\text{run}} = \dfrac{y_2 - y_1}{x_2 - x_1}$ where (x_1, y_1) and (x_2, y_2) are points on the line. Lines that go up to the right have positive slopes and down to the right have negative slopes. Horizontal lines have zero slope and vertical lines have no slope.

Slope-intercept form: A linear equation in the form of $y = mx + b$ where m is the slope of the line and b is the y-intercept.

Square: To multiply an expression by itself.

Square matrix: A matrix that has the same number of rows as columns.

Square root: The number that is multiplied by itself, or squared, to form a given number. Square roots are always positive (or zero).

Subset: Set A is a subset of set B is each element of A is also an element of B.

Substitution method: A method of solving a system of equations by using one equation to solve for a variable and substituting that expression into the second equation and solving.

System of equations: A set of two or more equations with two or more variables.

Term: Each monomial that is added to form a polynomial.

Transformation: Transformation takes the general shape of a function and either translates it (up, down, left or right), reflects it across an axis, or changes the scale of it. The general shape stays the same.

Trinomial: A polynomial with three terms.

Union of sets: Given two sets A and B, the set of elements that belong to either A or B or both. Written as $A \cup B$.

Universal set: A set from which other sets draw their members.

Variable: A letter or symbol that represents a quantity.

Venn diagram: A pictorial diagram showing logical relations between sets.

Vertical line: An equation in the form of $x = a$ creates a vertical line.

Vertical line test: A test to determine whether a relation is a function. If any vertical line through the graph intersects the graph in no more than one location, it is a function.

Whole number: A number from the set $\{0, 1, 2, 3, \ldots\}$.

x-axis symmetry: A graph which is a mirror image across the x-axis. In origin symmetry, points that are in quadrant I are reflected to quadrant IV and vice versa. Also, points that are in quadrant II are reflected to quadrant III and vice versa. Except for the line $y = 0$, relations with x-axis symmetry are not functions.

x-coordinate: The x-value in an ordered pair (x, y).

x-intercept: The x-coordinate where a graph intersects the x-axis.

y-axis symmetry: A graph which is a mirror image across the y-axis. In y-axis symmetry, points that are in quadrant I are reflected to quadrant II and vice versa. Also, points that are in quadrant III are reflected to quadrant IV and vice versa.

y-coordinate: The y-value in an ordered pair (x, y).

y-intercept: The y-coordinate where a graph intersects the y-axis.

Zero of a function: Values of x for which $f(x) = 0$. Also called solutions or roots of the equation.

Index

A

Absolute value
  equations, 123–124
  inequalities, 124–125
  of a number, 123
Additive inverse, 48
Algebraic expressions, 45–51
  associative property, 47
  commutative property, 46–47
  distributive property, 49–51
  evaluating the expression, 45
  formulas, important, 227
  identity property, 47
  inverse property, 48
  solving the problem, 231–232
Arithmetic means, 181–182
Arithmetic sequence, 178–180
  common difference, 178
  general formula for, 178–179
Arithmetic series, 182–184
Associative property of algebraic expressions, 47

B

Base, 24, 51, 145
Binomial, 57
Binomial theorem, 194–195
  coefficients of binomial expression, 195–196
  finding binomial expression, 198–202

C

Cartesian coordinate system, 81–82
CLEP College Algebra exam
  day of exam, 9–10
  options for military personnel and veterans, 7–8
  overview of, 5–7
  sample problems, 212–226
  for students with disabilities, 8
  test-taking tips, 10–11
  types of problems on, 209–212
  6-week study plan, 8–9
Closed dot value, 75
Closed set, 159
Closure concept, 159
Coefficient, 45
Combinations, 191–192
Common denominator (CD), 36
Common ratio, 184
Commutative property of algebraic expressions, 46–47
Complex conjugate, 171
Complex fractions, 70–71
Complex numbers
  addition of, 168–169
  definition of, 166–167
  division of, 170–171
  imaginary numbers, 165–166
  multiplication of, 169–170
  operations with, 168–172
  plotting of, 168
  solving quadratic equations with imaginary roots, 172–174
  subtraction of, 168–169
Composition of functions, 87–88
Conjugate, 71
  complex, 171
Consistent system, 135
Constants, 45
Conversion
  large numbers, 39
  scientific notation to standard notation, 38–39
  small numbers, 39–40
  standard notation to scientific notation, 39–40
Counting principle, 189
Cramer's rule, 206
Cross-multiplication, 74
Cubing of numbers, 25

D

Decay curve, 146
Degree of a polynomial equation, 108–109
Dependent system, 136
Diagnostic exam, 4
Discriminant, 131, 173
Disjoint sets, 18
Distributive property of algebraic expressions, 49–51
Domain of a function, 90–91

Domain of a relation, 81
Double inequalities, 78

E

Element, 203
Element of set, 15
Equations
   absolute value, 123–124
   advanced systems of, 143–144
   elimination method for solving, 139–142
   formulas, important, 229
   graphical method for solving, 137–138
   linear, 72–73, 94–96
   logarithmic, 153–155
   methods to solve a system of, 137–142
   quadratic, 125–133
   radical, 74–75
   rule for solving, 72
   second-degree, 112–113
   simultaneous, 135
   solutions of, 126
   solving the problem, 236–238
   substitution method for solving, 138–139
   third-degree, 113–114
   "trial-and-error" method to solve a system of, 137
Errors
   involving exponents, 243
   involving fractions, 241
   involving parentheses, 240
   involving radicals, 242
   miscellaneous, 244
Evaluating the expression, 45
Even functions, 101
Exponential functions, 145–148
Exponents, operations with, 24, 51–57
   multiplying and dividing expressions, 53
   negative exponents, 52
   for nonzero real numbers, 53
   positive whole number exponents, 51
   radicals and fractional, 56–57
   raising expressions with exponents to a power, 54–55
   of variables, 195
Extraneous solutions, 74

F

Factorials, 189–194
Factoring, 61
Factorization of polynomials, 65–66
Factors, 32

Finding a root, 26
   cube root, 26
   square root, 26
First-degree, functions, 91
F-O-I-L method, 58–61, 169, 194
Formulas, important
   algebra, 227
   equations and inequalities, 229
   functions, 228
   number systems and operations, 230
Fractional exponents, 56
Fractions, operations with, 32–37, 73–74
   adding/subtracting, rules for, 36
   cancellation of expressions, 33
   common mistakes with, 33
   dividing, rules for, 34
   in improper form, 32
   multiplication, rules for, 33–34
   reduced to lowest terms, 32
   reducing of fractions, 32
Full-length practice exams, 4–5
Functions, 84–91
   composition of, 87–88
   domain of a, 90–91
   evaluation of, 85
   even, 101
   exponential, 145–148
   first-degree, 91
   formulas, important, 228
   interval notation, 89–90
   inverse, 117–120
   linear, 91–101
   notation, 85
   numerical representation of, 85
   odd, 102
   piecewise, 88
   range of a, 91
   solving the problem, 233–236
   symbolically represented, 85
   verbal representation of, 85, 88–89

G

General form of an equation, 94
Geometric sequences, 184–186
   formula for the nth term of, 185–186
Geometric series, 186–187
Graphs, 81
   graphical representation, 85
   maximum number of real roots on, 111–113
   minimum number of real roots, 113–115
   plotting of graph equations, 82–84
   polynomial, 108–117

roots of functions on, 110–111
symmetry, 101–103
transformation of, 103–108
zeros on, 110
Growth curve, 146

H

Horizontal lines, 93

I

Identity property of algebraic expressions, 47
Imaginary roots, 116
Improper form of fractions, 32
Inconsistent system, 135
Index, 26
　of the radical, 56
Inequalities
　absolute value, 124–125
　definition, 75
　double, 78
　formulas, important, 229
　linear, 99
　quadratic, 133–134
　simple linear, 76
　solving the problem, 236–238
　symbols used with, 75
Infinite geometric series, 188–189
Integer coefficients, 61
Integers, 16, 160
　adding, rules for, 19–20
　dividing, rules for, 22–23
　multiplying, rules for, 21–22
　subtracting, rules for, 20–21
Intercept form, 96
Intersection of two sets, 18
Interval notation, 89–90
Inverse of a function, 117–120
Inverse property of algebraic expressions, 48
　additive inverse, 48
　multiplicative inverse, 48
Irrational numbers, 163

L

Least common multiple (LCM), 68
Linear equation representations of verbal
　　problems, 97–99
Linear equations, 72–73, 135–143
Linear function, 91–101
Linear inequalities, 99
Logarithms, 148–155
　common, 150–151

logarithmic equations, 153–155
natural, 151–152
rules for, 152–153
Lowest common denominator (LCD), 37, 68

M

Matrix (Matrices), 203–206
　determinant of, 205–206
　dimension of, 203–204
　scalar multiplication of, 204–205
　square, 203
Member of set, 15
Mixed number, 162
Monomial, 57
　degree of, 57
　multiplication of, 57–58
Multiplicative inverse, 48

N

Natural logarithms, 151–152
Natural numbers, 16, 159–160
Negative exponents, 52
Nested parentheses, 31
N factorial, 191
Nth root of a value b, 56
Number system, 16. *see also* real numbers
　cubing of, 25
　formulas, important, 230
　large, 39
　natural, 16
　rational, 32
　small, 39–40
　squaring of, 25
　substitute, 45
　whole, 16

O

Odd functions, 102
Online diagnostic exam, 4
Open dot value, 75
Operations
　with algebraic expressions, 45–51
　expanding and factoring polynomials,
　　57–66
　with exponents, 51–57
　with fractions, 32–37
　with integers, 19–23
　order of, 29–32
　with rational expressions, 66–72
　solving linear equations and inequalities, 72–78
Ordered pair, 81

Order of operations, 29–32
Origin, 82
Origin symmetry, 102

P

Parabola, 127
Parallel lines, 96–97
Parentheses
  nested, 31
Parentheses, use of, 21
Pascal's triangle, 196–198
Perfect cubes, 26
Perfect squares, 26
Permutation, 191–192
Perpendicular lines, 96–97
Piecewise function, 88
"Please Excuse My Dear Aunt Sally"
  mnemonic, 30
Point-slope form of line, 95
Polynomial
  graphs, 108–117
Polynomials
  definition of, 57
  division of, 60–61
  expanding and factoring, 57–66
  factoring of, 61–64
  factorization of, 65–66
  F-O-I-L method of multiplication, 58–61
Positive whole number exponents, 51
Powers, 24–26
Prime, 61

Q

Quadrants, 82
Quadratic equations, 125–133
Quadratic formula, 131–133
Quadratic inequalities, 133–134
Quotient, 22

R

Radical, 26, 56
Radical equations, operations with, 74–75
Radicand, 26, 56
Range of a function, 91
Range of a relation, 81
Rational expressions, 66–72
  addition and subtraction rule, 68–70
  cancellation across, 67
  definition, 66
  division of, 67–68
  multiplication of, 67

Rationalized denominator, 71
Rational numbers, 22, 32, 161–162
Real numbers
  definition of, 163–165
  formulas, important, 230
  integers, 160
  irrational numbers, 163
  mixed number, 162
  natural numbers, 159–160
  number sets, 159
  rational numbers, 161–162
  solving the problem, 238–240
  whole numbers, 160
REA Study Center, types of assessment at, 4
Reciprocal, 34
Relation, 81
  domain of a, 81
  range of a, 81
Remainder theorem, 61
Roots, 26–29, 110
  eliminating of, in denominator, 71–72
  finding a root, 26
  finding the cube root, 26
  finding the square root, 26
  imaginary, 116
  maximum number of real, on graph, 111–113
  minimum number of real, on graph, 113–115
  multiplicity of, 114
  nth root of a value b, 56

S

Scientific notation, 37
  conversion of standard notation to, 39–40
  conversion to standard notation, 38–39
  multiplication and division of numbers in, 40–41
Sequences
  arithmetic, 178–180
  general, 174
  geometric, 184–186
  infinite, 174
  nth term or general term of, 174–175
  term of, 174
Series
  arithmetic, 182–184
  definition of, 176
  general, 176–177
  geometric, 186–187
  infinite geometric, 188–189
Set (s)
  complement of A, 17
  definition, 15–16

disjoint, 18–19
element, 15
empty, 16
finite, 16
infinite, 16
member, 15
of numbers, 16
subset, 16–17
union and intersection of, 17–18
Sigma notation, 177
Simultaneous equations, 135
Slope-intercept form, 94–95
Slope of a line, 92
Solutions of the equation, 126
Square matrix, 203
Squaring of numbers, 25
Subsets, 16–17
Substitute, 45
Symmetry, 101–103
System of linear equations, 135–143

T

Terms, 45
Test-taking tips, 10–11, 245–246
Trinomial, 57

U

Union of two sets, 17–18

V

Variables, 45
Venn diagram, 17
Verbal representation of function, 88–89
Vertical lines, 93
Vertical line test, 85, 109

W

Whole numbers, 16–17, 160

X

X-axis, 81
X-axis symmetry, 102

Y

Y-axis, 82
Y-axis symmetry, 101
Y-intercept, 94